Praise for *Spoonfuls of Honey*

'There's a buzz around the magic of bees and honey, and if anyone is in tune with it, it's the award-winning writer Hattie Ellis... Packs in facts and useful tips.' *delicious.*

'Hattie Ellis' thoughtful account of the bee and the history of honey is matched with 80 honey-inspired recipes.' *Telegraph*

'Really draws the reader in to the world of all things honey.' Shortlisted for the Guild of Food Writers Awards Best Cookery Book, 2014

Praise for *Sweetness & Light: The Mysterious History of the Honeybee*

'Richly informative and beautifully written.'
The Times

'An evocative work that feels, smells, and tastes like everything to do with bees. Readers will be absorbed into the bee yard with her as she explores images and sensations of beekeeping with all senses alert.'
New York Sun

Spoonfuls of
Honey

Spoonfuls of Honey

Recipes from around the world

Hattie Ellis

PAVILION

For Willie, Daphne, Stephen, Heather and
Frances, and the next William, with love and
thanks for all you do at Chain Bridge Farm

This edition published in the
United Kingdom in 2019 by
Pavilion Books
43 Great Ormond Street
London WC1N 3HZ

ISBN: 978-1-911624-70-7

10 9 8 7 6 5 4 3 2 1

A CIP catalogue record for this book is available
from the British Library

www.pavilionbooks.com

Reproduction by Rival Colour Ltd, UK
Printed by Toppan Leefung Printing, Ltd

Notes
Honey is not recommended for children under
12 months because of the risk of botulism
Medium eggs are used unless otherwise stated.
1 teaspoon (tsp) = 5ml; 1 tablespoon (tbsp) = 15ml
All spoon measurements are level

CONTENTS

WHAT IS HONEY?

Honey has the sweetest associations. Put a dab on your tongue and let its smooth sugars dissolve into a long hit of flavour and energy. It makes you think of summer days when bees buzz between flowers in the sunshine.

This benevolence spreads into the kitchen. A teaspoon of honey sweetens and deepens a tisane or stew and adds lustre to a sauce. Sweet tarts, cakes and roasts are burnished by its glow. Syrup-drenched baklava, glazed chicken wings and sticky ribs are made special with a touch of honey. Glistening threads are drizzled over crisp fritters or slices of Italian cheese such as pecorino. A spoonful of honey keeps breads soft and fresh; cakes and gingerbreads made with spices and honey improve as the moist crumbs soften and the flavours meld over time. You don't need much honey in a dish, but it always makes a difference.

There are many varieties of honey to explore. Its colours range from palest gold to deep mahogany, its flavours from gentle sweetness to a fragrant bitterness. Honey comes from wild flowers, woodlands, urban gardens and rainforests. It gives you a sense of place, be it a London park, Scottish moor, Tuscan woodland, Alpine meadow, Australian forest or Californian mountainside.

For all this variety, we should start at the beginning. What is honey?

Honey is concentrated nectar, pure and simple. Nectar is a sugary liquid, around 70–80 per cent water, which is collected from plants by the bees and carried back to the hive in a small internal sac known as a 'honey tummy'. Back in the dark interior of the hive, the nectar is reduced right down as thousands of honeybees pass droplets from one to the other and fan their wings to evaporate the water. It reduces down to a sticky substance that is around 18 per cent water and is stored in the little hexagonal wax cells of the honeycomb, as if in small pots. The bees cover each cell with a wax cap.

What is this honey for? It is food for the bees, their carbohydrate source to give them energy to fly and collect more nectar and make more honey. They collect this sticky liquid energy in the long warm days when nectar flows and store it in combs to use when needed.

The honey in our pots has been extracted from honeycomb by beekeepers and stored so that we, too, have sustenance on hand. It was mankind's first and most potent form of sweetness, predating the widespread use of sugar by many millennia. Long before hives were invented, honey hunters sought out caches of sweet comb in the wild.

From prehistory onwards, mankind's love of bees, honey and all that they symbolize has been part of music, poetry, architecture, art, philosophy, politics and religion. We yearned for a Biblical land of milk and honey; we see the hive's community as a

metaphor for human society and we design buildings like honeycomb; we sing of the bee's buzz. People all over the world have used honey for ceremony and celebration. Fermented honey, or mead, was one of the first means of intoxication, long before grapes were pressed to make wine. Wax from the comb provided an early form of light, the sweetly scented candles that lit churches and wealthy homes. Throughout human history honey has seemed so beautiful and good as to be almost miraculous, and honeybees were seen as mysterious creatures of supernatural power.

Honey is just as significant, in these days of urban life and environmental uncertainty, through its closeness to nature. Appreciating this gives the honey-lover a more intimate and knowledgeable connection with the natural world around us, to be found even in green corners of the inner city.

Inspiration for the recipes in this book comes from a sense of the natural history of food, seen at close quarters. Most days in the spring, summer and early autumn you can go outside and watch bees land on blooms, bury their heads in the centre as they seek out the nectar, and come away laden, their bodies dusty with pollen. Then you can go back to the kitchen, put a spoon in a pot and taste this connection.

A number of the recipes come from the many countries I've visited to learn about food and talk to beekeepers, from New Zealand to New York City. Many of the recipes use just a small amount of honey. Bees gather nectar from some two million flowers to make a single 450g/1lb jar of honey. Honey is powerful and precious. This book will show how to use it to best advantage and enjoy it most; an appreciation of honey, bees and the sweet goodness of one of the most extraordinary foods in the world.

A—Z OF HONEY

When peering into the golden depths of a pot of honey at home, choosing what to buy, or talking to honey producers, a little knowledge can be a useful thing.

BLENDED HONEY

The most widely available commercial honey is blended from various sources by honey packers in order to get a consistent product for different retailers. This sort of honey is less expensive than pots produced by beekeepers from their own hives, which contain honey made from the nectar of flowers of a particular time and place.

COLOUR

Honeys vary in colour from very pale gold – the lightest category in the United States is called 'water-white' – to dark brown-black. The colour (and flavour) of the honey depends on the nectar of the plants the bees visit. There is the pale, greeny gold of apple blossom honey, the exquisite light gold of acacia honey, the foxy red of heather honey, the amber of manuka honey and the near-black of some forest honeys. Darker honeys contain more minerals and are generally slightly more nutritious and stronger-tasting than lighter ones.

CRYSTALLIZATION

The texture of honey depends partly upon the composition of sugars in the nectar, and partly upon the age of the honey and how it is produced and stored. Some honeys crystallize quickly, while others remain runny for longer. Honeys with a high proportion of fructose, such as acacia, will stay liquid for much longer than those with more glucose, such as oilseed rape honey. *See also* Set, creamed or whipped honey.

FILTRATION

Once the honey has been removed from the comb it is usually filtered. Smaller-scale beekeepers may simply strain their honey to remove insect legs, pieces of wax and so forth, or leave it to settle in small containers so the wax floats to the top and the clean honey is drawn off the bottom, the method used for 'raw' or artisanal honeys; pollen grains remain in the honey, often giving it a slightly hazy appearance. More commercial honey is heated to varying degrees and filtered through a finer mesh, which removes more of the pollen grains to make a honey that is clearer, smoother in texture and slower to crystallize.

FOREST HONEY *see* HONEYDEW HONEY

FRUCTOSE *see* SUGARS

'FUNNY HONEY'

Honey that is not what it purports to be is known in the trade as 'funny honey'. It may be honey mixed with corn syrup or another form of cheap sugar. It may be a cheap imported commodity honey repackaged as a local or special honey. It may be illegally imported from a country whose honey exports have been banned. Or it may be ultra-filtered honey, which is diluted, passed through a fine mesh and evaporated to the right consistency and so highly processed that it should no longer be called honey.

GRANULATION *see* CRYSTALLIZATION

HONEYCOMB

Honeycomb hangs vertically in the hive, with the hexagonal cells angled to help hold the honey within. Beekeepers sometimes sell comb honey, for some the ultimate form of honey as it contains simply the bees' food, as nature intended, and at its most fragrant and delicious. The wax is indigestible but you can either swallow it or remove the ball of wax that forms in your mouth as you eat.

HONEYDEW HONEY

Most honey comes from nectar, but honeydew or forest honey comes from the sweet secretions of aphids and other insects that feed on tree sap. The bees produce a dark, mineral-tasting honey that may be labelled pine, oak or beech or forest honey and sometimes mountain honey. Honeydew is a love-it-or-hate-it honey; for fans, it is an unusual honey with a strong appeal.

MANUKA AND OTHER 'HEALTHY' HONEY

Folk medicine has long used honey for health, and certain honeys were and are regarded as especially beneficial. New Zealand manuka honey was the first honey to be proven to have particularly strong antimicrobial properties. Batches of manuka honey are laboratory-tested in order to be given a grading of a 'unique manuka factor' (UMF) and other grading systems to show their potency. A number of other types of honey, such as jarrah, have also undergone tests to show their efficacy in order to be marketed as healthy honeys. See Honey and health, *page* 33, and Around the world in 90 pots, *page* 182.

MEAD

Fermented honey makes an alcoholic drink. Mead can be sweet or dry and served as an aperitif or dessert wine, complementing such foods as cheese, pâté and desserts. Spiced mead, which is flavoured with ingredients including mace and cinnamon, is called metheglin, while mead made with honey and pressed fruit juice is called melomel. d is an ancient drink, celebrated in many cultures and featured in poetry such as *Beowulf* and *The Canterbury Tales* and in Shakespeare. Commercial 'mead' may be sweetened poor-quality wine. Real mead, however, is a drink worth rediscovering and is undergoing a revival. The best mead producers use high-class honey and the very best mead is aged, as it improves with time.

MONOFLORAL OR VARIETAL HONEY

Honey that comes largely from a single nectar source is called monofloral or varietal. Honeybees will return to a good nectar source and so produce a particular kind of honey with a distinctive character. Bees generally gather honey within half a mile or a mile of their hive, and up to around three miles, going to the nearest and best available nectar source. Placing the hives in the right place, such as on a heather moor, enables the beekeeper to produce a varietal honey. Beekeepers who move their hives to be close to a commercial crop that needs pollinating (such as oranges) get a varietal honey almost as a by-product of this process. Monofloral honeys will contain nectar from other plants but should be predominantly of one kind.

MOUNTAIN HONEY *see* HONEYDEW HONEY

MULTIFLORAL HONEY

Most honey comes from the nectar of a variety of plants. Even monofloral honeys will contain nectars from other plants, sometimes in quite large amounts. Multifloral honeys that come from a particular place can have a very strong character and be as distinctive and delicious as single-varietal honeys. For instance, Greek mountainside honey tends to be predominantly from the nectar of marjoram, thyme and other mountain herbs, and wildflower honey includes the flora of a particular area, e.g the Alps.

NUTRITION

Honey is mostly composed of sugars (fructose and glucose) with 17–19 per cent water and 0.5 per cent a complex mixture of antioxidant vitamins, minerals and enzymes. The sugars give you energy and the other elements, although small, make honey more nutritious than table sugar. Darker honeys in particular, such as buckwheat and manuka, tend to have higher levels of antioxidants and other nutritional benefits. The enzymes and other beneficial phytonutrients are most active in honey that hasn't been heated or over-processed.

Gram for gram, honey is lower in calories than table sugar (sucrose), but because it is denser, honey has 64 calories per tablespoon, while a tablespoon of sugar has around 48 calories. However, the high amount of fructose in honey means it tastes sweeter than table sugar, and combined with its delicious flavour, a little goes a long way.

ORGANIC HONEY

The definition of organic honey is that the bees are kept on land that is farmed organically, without the use of synthetic pesticides or fertilizers. In Europe this must be 3km/1¾ miles of land around the hive; in Britain, honey certified by the Soil

Association is even more strictly regulated: the hives must be surrounded by 6.4km/4 miles of organic forage.

RAW OR UNPASTEURIZED HONEY

These are terms used by artisanal beekeepers, generally meaning that their honey isn't heated at all, or not heated above the normal hive temperatures of 33-35°C (91-95°F), in order to best capture honey's natural goodness and tastes.

Because it is only coarse-filtered or strained, rather than being fine-filtered, raw honey will contain more pollen than filtered honey – and perhaps small pieces of wax – and may be slightly hazy. It may also crystallize at a quicker rate than processed honey.

SCENT

Depending on its nectar sources, good honey can be highly scented, especially when it is freshly harvested. Lavender and orange blossom are examples of honeys that are redolent of their source, but honey doesn't necessarily carry the scent of the flower it comes from. The volatile compounds in the honey can create a wide range of aromas (and tastes), from floral to vegetal, woody, burnt sugar and spices (*see* How to taste honey, *page* 29).

SET, CREAMED OR WHIPPED HONEY

In its natural state, fresh from the comb, honey is runny. To suit some consumers, runny honey can be artificially crystallized by stirring in a small amount of finely crystallized honey, known as a 'seed' honey; as the crystals spread throughout the honey, it acquires a fine, firm, even texture. Seeding makes a honey lighter in colour and less aromatic.

SUGARS

Honey is mostly made of two types of sugar, fructose and glucose. Honeys with a higher proportion of glucose tend to crystallize more quickly. Fructose tastes sweeter than table sugar (sucrose), which means you don't need as much to get the same effect.

WORLD TRADE

Honey is both a local food and a world commodity. China, the European Union, Turkey, Ukraine, Argentina, the United States and Mexico are the largest producers of honey on the world market. Packers import honey in bulk from different countries and blend them to make a consistent product for retailers. Specialist honey retailers seek out pots from smaller producers, both locally and from other countries, who produce especially tasty honeys, such as tawari from New Zealand, chestnut from Italy and Spain and heather from Britain (*see* Around the world in 90 pots, *page* 174). Honey for export has to be cleared to ensure it doesn't spread bee disease.

VARIETAL *see* MONOFLORAL

A—Z OF HONEYBEES

BEES

Of the 25,000 kinds of bees, the *Apis*, or honeybee, genus is the one best known for producing honey. The most successful honeybee of all at this task is *Apis mellifera* ('honey-bearer'), originally from the Middle East and Africa and now found around the world, including in the United States, Australia and New Zealand, where it was taken by Europeans. The honeybee is a social insect and lives and works in a colony that stores honey in harvestable combs.

BEEKEEPING

Beekeepers provide hives to house colonies of honeybees; in the summer a hive may contain around 60,000 insects. Due to problems with diseases, especially since the spread of the parasitic varroa mite, it can be hard for honeybees to survive in the wild; they now live largely in man-made hives that are checked and treated for disease. As well as harvesting honey, beekeepers keep honeybees alive. The beekeeper puts boxes, or 'supers', on top of the part of the hive where the bees produce the brood of new bees. The bees build honeycomb in these supers and these are removed by the beekeeper to extract the honey.

BEESWAX

Malleable wax is secreted by the bees and moulded to form the hexagonal cells of honeycomb that are used to store honey. After the honey has been removed, the wax can be melted down to make candles that carry the sweet scent of the hive; other uses include making polish for furniture and leather.

COLONY

A colony of bees consists of a single queen, a few hundred male drones and tens of thousands of female worker bees. The queen is the largest bee in the colony and lays as many as 2,000 eggs in one day. The drones exist to mate with the queen on her virgin flight out of the hive. The workers are the heroes of the hive, the ultimate multi-taskers whose collective power keeps the colony going. They do everything from feeding the developing larvae in the brood comb to foraging for nectar and pollen and defending the entrance of the hive from invading insects. All the bees you see flying around flowers are worker bees.

HIVE

A beehive is essentially a box in which the bees build their comb, just as they do in nature by finding a hollow in a tree. Ancient Egyptian art shows images of beekeeping using simple hives. The old-fashioned skep, the traditional domed beehive made of straw, was an early form of hive, but the bees had to be destroyed to harvest the honey. These were superseded in the nineteenth century by modern removable frame hives that allow the extraction of frames of honeycomb while leaving the bee colony to breed and produce more honey.

HONEY HUNTING

In the days before beekeeping, and still in some cultures today, honey hunters tracked down wild honeycomb and endure stings to seize their sweet prize. In Africa, some hunters follow honey-hunting animals such as the honeyguide, a small bird that is unusual in that it can digest beeswax. The honeyguide deliberately leads humans to a wild bees' nest by calling, flying short distances and then stopping to call again. Once at the nest, the human hacks open the tree containing the honeycomb, or digs it out from a hole in the ground, and gives the bird its sweet reward.

NECTAR

Honey is concentrated nectar, the sweet substance that attracts bees to flowers. The sugar content and quantity of nectar varies enormously from flower to flower. A clover floret may produce one-twentieth of a pin's head of nectar, while other flowers, such as some of the Australian eucalyptuses, flow briefly but profusely with nectar, although not every year. Some flowers produce nectar at a certain time of day and only above a certain temperature. Bees go to where they can get the most nectar at any particular time and work those flowers as much as possible.

POLLEN

Pollen is the bees' source of protein and also contains vitamins and minerals; it is sold as a health food for humans. The pollen is collected by the beekeeper in a small pollen trap that is put on the entrance of the hive for a short time to brush off and collect the

pollen balls from the bees' legs. The pollen in less processed or 'raw' honey gives it a slightly hazy appearance.

POLLINATION

Bees and flowering plants evolved at the same time in the Cretaceous period, between 140 and 60 million years ago. Bees and blooms need each other: the bee feeds on the bloom's nectar and pollen; the flower needs a creature to transfer the male reproductive cells, in the pollen, to the female parts of the plant in order to produce its seeds, nuts or fruit. Many human food plants rely on insect pollination. Ants, other bees and other creatures are also important for pollination, but honeybees are especially numerous and can be organized in hives, so are particularly useful for this task. Honeybees can boost crop yields by 60 per cent.

PROPOLIS

Alternative health practitioners and beekeepers alike are big believers in the power of propolis, the resin exuded by trees and other plants to seal their wounds and collected by bees to plug gaps in the hive. A beekeeper scrapes a little propolis straight off the hive to chew to ward off a cold, and dentists use its antiseptic properties for oral health. Propolis is available in tinctures and other forms.

ROYAL JELLY

This milky white gelatinous substance is secreted by worker bees and fed to the developing larvae in the hive. Those fed for a few days, and then fed a mixture of pollen and nectar called 'beebread', become female worker bees or male drones. Protein-rich royal jelly is the sole food given to certain larvae in larger cells that then become the queens. Because a queen bee may live for several years, as opposed to the average summertime worker's life of around six or seven weeks, royal jelly is regarded as promoting health and longevity and is used as a health supplement, especially in Japan and other countries in East Asia.

HONEY IN THE KITCHEN

Honey is one of the most delightful — and versatile — pots on the shelf. Whether you use it as an ingredient, a seasoning or to add the finishing touch to a dish, honey finds a place at every kind of meal.

Savoury as well as sweet foods are enhanced by honey. Its flavour notes add depth to many dishes, from stews and tagines to soups, roasts, savoury pastries, salads and sauces. As for sweet dishes, these are given an entirely different dimension when you replace some or all of the sugar with different kinds of honey.

To develop the recipes in this book, I talked to cooks, chefs and beekeepers and scoured every kind of cookbook, both old and new. Honey was the earliest form of sweetness in kitchens all over the world, predating the common use of sugar. Ancient Roman and Greek and medieval Arabic cooking used honey in dishes that are still enjoyed today, such as baklava and honey-gingerbreads.

Honey is a time-honoured ingredient, but is also strikingly modern. There's now a greater awareness of the natural world in cooking, with the use of botanicals, flowers, wild foods and honey in recipes. A decade ago, honey may have been seen as a touch quaint or 'country-cute'; now it is bang-on contemporary.

Just as honeybees are found all over the world, so are recipes that use their honey. It combines with many spice blends and ingredients from different cuisines in a way that is infinitely varied. Asian flavours such as five-spice powder work brilliantly with honey. Jamaican jerk, Spanish tapas, an English fool, a French sauce, an American muffin, a Brazilian salad: all are enhanced by honey. Explore honey and you travel the world. Many of the tricks using honey are transferable to other recipes. Moroccan cooks, for example, often add a touch of honey to tomato-based sauces to soften their acidity and round out their sweetness. You can use this idea in pasta sauces and other tomato sauces or in a dressing for tomato salad, especially if the tomatoes are not the ripest.

A further source of inspiration for using honey comes from being outdoors. I've long been a bee-spotter, watching blooms nod as the bees gather nectar and pollen. The association of bees with herbs and edible flowers suggests many combinations of flavours and colours. Most of all, ideas come from having honey pots sitting in your kitchen. You soon find ways to make them part of your cooking.

HOW TO MAKE HONEY RUNNY

Recipes sometimes call for 'runny honey'. Much honey is fine-filtered so that it remains runny. Some artisanal honeys, depending on their nectar sources and how they are kept, will crystallize and harden naturally as their sugars set, but it is easy to make them runny again. Put the whole jar — with the lid slightly loosened — in a small bowl or jug and pour in just-boiled water so it surrounds the jar. Leave it for 15 minutes or so, giving the honey a stir towards the end.

If you want to quickly get a spoonful or two of liquid honey, for example in order to brush it onto meat as a glaze or so it melts more easily into a sauce that won't be heated, such as mayonnaise, you can put a couple of tablespoons in a bowl in the microwave and heat the honey on low power for about 10 seconds. Take care not to overheat the honey; it takes no time at all to make it runnier. If you don't want to put good honey in a microwave you can put the spoonfuls in a bowl over a small pan of boiling water and let the heat melt the honey in seconds. All this is a slight faff, however, and wastes some of the honey, and so I tend to use honey that is already runny when using just a little in this way.

Often the heat of the dish is enough to melt the crystals. To glaze a piece of meat hot from the oven, spread or spoon on the honey and it dissolves instantly. The crystals will also melt in a cooler liquid, such as a salad dressing, if left for a short time.

MEASURING HONEY

Honey is sticky stuff, but there are several simple tricks to make it easier to handle and measure. The most universally useful method is to dip the measuring spoon first into very hot water. The honey will then slide easily off the hot spoon. Alternatively, when you are cooking a dish that uses oil, measure out the oil with a spoon and use that spoon to measure out the honey. The honey will slip easily off the spoon and into the pan or bowl. Whatever method you use, it is useful to have two spoons on hand when measuring honey: one for measuring and another for scraping off the last drops.

When you need a large quantity of honey it is best to measure it directly in the bowl or pan that you are going to use, as this means less waste. Place the bowl or pan on the scales and reset the weight to zero, then add your honey. Go carefully as honey is heavy and just a small amount quickly adds up the grams.

CHOOSING THE BEST HONEY FOR A DISH

Any honey will suit any dish in this book. There is an essential honey flavour that will sweeten and smooth out a dish and make it shine. All you need is a single pot of inexpensive honey and you are in business. But once you get into the subject, curiosity and pleasure will no doubt lead you to try out different kinds.

There are three broad categories of honey, described by colour: light, medium and dark, and the recipes sometimes suggest when one of these would work best, although any honey will do.

Light honeys, such as acacia or orange blossom, are great for subtle dishes, for example for adding to berries or a mild goat's cheese. Medium honeys have a bit more oomph and might be used with plums, stronger cheeses such as Cheddar, or meat. These two types, light and medium, are broadly interchangeable. The darker honeys, such as chestnut and buckwheat, have a more insistent taste that can be strongly resinous or malty and are best used judiciously and in balance with other strong flavours. Spiced cakes, such as a rich fruit cake, pain d'épices or gingerbread, work well with a stronger-flavoured honey, as do spicy marinades and glazes, such as jerk chicken, spare ribs and barbecue sauce.

While stronger honeys can be a touch overbearing if you're not careful, a small amount of a stronger honey spreads its flavour through a dish. Madeleines (*see page* 120), those delectable French honey buns, use just a little honey. I find that heather honey gives them a more distinctive flavour than a lighter honey such as orange blossom. But stronger honeys are a matter of taste; some people don't like their tang, whereas everyone loves milder honeys such as acacia or clover.

It is good to have a small range of honeys within sight. Jars of honey add a golden glow to the kitchen that suggests light, warmth and flavour. The glow catches your eye and a spoonful of honey in a dish will often combine with other ingredients in a way that surprises and delights.

MAKING THE MOST OF SPECIAL HONEYS

Each recipe in this book is designed to give 'value for honey': each spoonful – sometimes just a single spoonful – will play its role in a recipe. Just as a pinch of salt or sugar brings other flavours to life, so a little honey works wonders, bringing a warm depth of flavour and aromatic qualities.

On the whole, it makes sense to use expensive and unusual honeys in dishes that do not involve heating or cooking the honey. This makes the most of the subtleties in their flavour and aroma. It is especially true of light or medium honeys, whose fragrance and flavour may be lost if heated.

When you have special pots of honey, the temptation is to keep and not to eat. While honey keeps well, its aromatic qualities do lessen over time. 'Little and often' is my policy with good honey. It is best to use small spoonfuls in many dishes while the honey is at its very best.

Some varietal honeys have a particular resonance with other flavours and it is fun to play around with this. This is especially true if you can use a honey from the country that is the origin of the dish. Greek mountainside honey, made from nectar gathered from wild marjoram, thyme and other herbs, fits beautifully into Mediterranean-style dishes, such as Chicken with honey, lemon and thyme (*see page* 65) or Goat's cheese, herb and honey filo parcels (*see page* 80). Lavender honey lemonade (*see page* 170), while already flavoured with flowers, somehow tastes all the better for using lavender honey as well.

HONEY IN BAKING

Honey works brilliantly in baking, adding flavour as well as sweetness. The general rule of thumb is that you can replace up to half the weight of sugar in a recipe with honey. I don't usually replace quite as much honey as that, in order to use honey more frugally, and often put it together with a light brown sugar.

The sugars in honey, especially fructose, taste sweeter than sucrose (standard sugar), so you can use less honey than sugar to get the same effect, cutting the total amount of sugar by about a quarter.

Be careful that the honey doesn't lead to a burnt bake. Honey browns faster than sugar, and I've adjusted temperatures and cooking times so that this shouldn't happen. On the whole, if replacing sugar with honey you need to reduce the oven temperature by 25°C/75°F. Ovens vary and many have 'hot spots', where food cooks quickly. If in doubt, turn a cake or baking tray around halfway through cooking to ensure it gets an even heat and keep a close eye on how well the bake is browning towards the end. Cover the top with foil if it's getting too dark. If using a fan oven, you may need to adjust the temperature to 10 or even 20 degrees lower than that given in the recipe; it all depends upon how hot your oven gets.

As well as adding flavour, honey improves the keeping quality of a cake, bread or biscuit. It is hygroscopic, meaning it attracts moisture and so stops the bake from drying out, which is why it is used in commercial baking. The messy remnants of honey from the comb is often referred to as 'baker's honey' because this is how it is used. This moisture-attracting quality means that, while honey cakes tend to be moist, honey biscuits aren't always super-crisp. Honey bakes are often best eaten a day after cooking, when the honey flavour will have developed slightly. Keep honey biscuits and cakes in an airtight tin and wrap the cakes in greaseproof paper too.

GLAZING AND DRIZZLING

The visual impact of honey gleaming in a glaze or drizzled over food is one of its most appealing qualities. Nothing gives quite the same glow.

A honey drizzle stick is a good-looking tool, but tends to hold the honey, and this can lead to waste, unless you keep the stick permanently in the honey pot. I tend to use a hot spoon, so most of the honey slips off easily as you drizzle.

A glaze imparts sweetness and flavour to a dish, as well as looking good. What's crucial is that you burnish rather than burn. In order to do this, I often add the honey towards the end of the cooking time. As well as keeping more of the honey's flavour intact, this means there is less risk of its sugars burning. In some dishes, such as Chicken with honey, lemon and thyme (*see page* 65), you want this almost-charred flavour. But I veer towards cooking honey less rather than more and sometimes add a bit of liquid alongside the honey so it doesn't burn so quickly.

Drizzling and glazing are best done with runny honey. If you have a more viscous or crystallized honey, simply heat it gently and briefly and the honey will be easier to brush on or drizzle. However, when glazing a hot piece of meat or fish, the heat of the food quickly melts even hard honey. Glazing a tart or cake with honey means you add less sugar to begin with, knowing that sweetness as well as a shine will be added at the end.

HONEY'S BEST FRIENDS

Particular foods work especially well in partnership with honey. Some combinations have centuries of tradition behind them and are being rediscovered by contemporary cooks and chefs.

Cheese, cream and other dairy products

All dairy products – from cows, sheep or goats – work brilliantly with honey. Cheese and honey have been put together since the Biblical land of milk and honey and the very first cookbooks by the ancient Greeks and Romans, with their honeyed cheesecakes.

The classic Italian pairing of pecorino or other aged sheeps' cheese drizzled with thyme, mountainside herb or chestnut honey is just one of an infinite number of happy combinations. When you want to showcase a special honey, go back to the beauty and simplicity of a single good cheese drizzled with honey, with perhaps a few nuts and some apple, pear or other fruit alongside. The balance of sweetness and acidity in both cheese and honey (and fruit for that matter) makes this a special and interesting dish. Goats' milk cheeses, with their lactic acidity, are especially good with honey.

Another reason this partnership works is that globules of dairy fat spread the sweet taste of honey gently throughout your mouth. Bread and honey go so well together partly because of the butter on the bread. Yogurt and honey is a classic combination on the same principle.

Goat's cheese, herb and honey filo parcels (*see page* 80) and Grilled goat's cheese and chicory salad with honey and poppy seed dressing (*see page* 82) are two savoury dishes that play on the classic pairing of cheese and honey. Ricotta hotcakes (*see page* 44) are a satisfying breakfast dish to eat with honey. Ice creams such as Chocolate and chestnut honey ice cream (*see page* 156) and an Indian Honey and cardamom kulfi (*see page* 154) use cream to good effect, and the Honey and whisky truffles (*see page*163) work so well because of the cream and butter in the ganache.

Fruit

The essential connection between honey and fruit makes them good partners on the plate. Bees play a vital role in the pollination of plants to make fruit, and fruit trees' nectar becomes honey in the pot. Orange, lemon and mango trees are just three of the fruit trees that produce plenty of nectar for honeybees to make delectably flavoured honeys. This connection is explicitly celebrated in Pollinators' fruit salad (*see page* 137).

Fruit and honey can also be used in savoury dishes, for example in the Roast grouse with a honey, blackberry and whisky sauce (*see page* 63) with its beautiful pink blackberry and honey gravy.

Nuts

Honey and nuts are another time-honoured duo. Middle Eastern sweets described in medieval cookbooks, and still made today, include many combinations of pine nuts, almonds and walnuts with the copious honey of this region. Honey and nuts feature in every chapter of this book, from Baked cheese with honey-walnut toasts (*see page* 79) to Smoky honey almonds (*see page* 107) to nibble with drinks and Turkish pinenut, yogurt and orange cake (*see page* 113). Just as bees and honey are connected to fruit, so they are to nuts. Billions of honeybees, in 1.5 million hives, are taken every year to the Central Valley in California to pollinate the almond blossom and produce nuts.

Herbs

Herb gardens are alive with bees. Marjoram is the sweetest and most attractive nectar for bees, while mint and thyme are other herbs that yield especially good and fragrant nectars which make delicious honeys. I like to put basil with the Honey-buttered French peaches (*see page* 143) and to use thyme with apricots (*see page* 53). The resinous tang of rosemary is the perfect partner for the almost medicinal nature of manuka honey in Manuka, rosemary and orange cordial (*see page* 169). Soft herbs are central to Honey sauce vierge (*see page* 100). Herbs pattern and flavour the honey jelly on top of Chicken liver parfait (*see page* 66) and make beautiful tisanes that can also be turned into such wicked drinks as the Moroccan mint tea cocktail (*see page* 173).

Spices

The warmth of honey is a good partner for spices and its sweetness mellows and helps meld together different spice combinations and to carry their flavours through a dish. Playing around with spices is one of the most creative areas of cooking; the taste equivalent of paint pots for an artist. The spice mixtures in Honeyed chicken and aubergine biryani (*see page* 68) and Lamb and honeyed apricot tagine (*see page* 59) work even better because of the honey. Spices can have heat as well as fragrance, and honey can smoothe out a fiery ingredient, for example in Sweet 'n' hot jerk chicken wings (*see page* 72) or the mild but buzzy pink peppercorns in the Smoked fish and honeyed potato salad (*see page* 74).

Some spices have a special affinity with honey. Saffron is one, and stars in Gooseberry, honey and saffron fool (*see page* 138). Cardamom and vanilla are also much used with honey. The Teatime baking chapter (*see page* 108) contains many traditional bakes that are also heady with the likes of cinnamon, nutmeg and allspice.

Flowers

Bees fly between blooms in gardens, parks and the countryside. Flowers instantly make a plate look special, but they are far more than mere decoration; as an ingredient, flowers work brilliantly with honey. One of the most famous of all honeys comes from lavender flowers and these are used to infuse Lavender honey lemonade (*see page* 170). Edible flowers are used in the Blue cheese and flower salad with honeyed walnuts (*see page* 76), and in Elderflower fritters (*see page* 134).

HONEYCOMB

Honeycomb is made up of hundreds of small wax cells in which the bees store their honey. You quite often see chunks of honeycomb in pots and more occasionally larger slabs of honeycomb or small square sections in a box. It is more expensive than honey in pots but worth every penny. While you do get chewy balls of wax in your mouth when eating honeycomb, it is a special treat because it is so natural and beautiful.

Honeycomb is a gift to the cook because all you need to do is slice off thin pieces and put them on a plate alongside a simple accompaniment. A slice of honeycomb, a fig and a blob of good cream or gentle cheese makes a beautiful dessert. Use a hot knife to slice off the comb to get a neater finish. Buttermilk pannacotta with honeycomb (*see page* 144) is a slightly more elaborate way of showcasing honeycomb.

POLLEN

Pollen is commonly found in health food shops and specialist honey shops. Its colours in nature vary, but commercially sold pollen is generally yellow-orange. The sweetish and slightly grainy texture of pollen make it an unusual ingredient that is fun to explore. It is often added to porridge and breakfast cereals to make them extra-fortifying, or used to add a final flourish to the look and taste of a dish, be it a soup, salad, bake, pudding or drink.

I have used pollen in the Muhammad Ali smoothie (*see page* 51), Honey-caramelized fennel soup (*see page* 84), Leeks with spicy pollen breadcrumbs (*see page* 96) and Pollen butter shortbread (*see page* 125). All these dishes can be made without pollen, but once you have a packet it's handy to have specific ideas about how to use it.

MEAD AND OTHER HONEY DRINKS

The sugars in honey ferment to make the alcoholic drink mead and its cousins: melomel, made from honey and fruit juice, and metheglin, flavoured with spices such as mace and cinnamon. While mead's good name has been tarnished by sickly imitations that just add honey to poor wine, there has been a revival of quality meads. These work beautifully as an accompaniment to puddings, rich pâtés or blue cheese. As an ingredient, mead adds a luscious sweetness to gravies and sauces, and a beautiful honey flavour to poached fruits, as in Pears poached in mead (*see page* 148).

A number of award-winning brewers are rediscovering the use of honey in beers. It also appears in spirits, from Polish honey vodka to Brazilian cachaça, and in liqueurs such as Drambuie. As with mead, these drinks can be added to dishes to give them a touch of honey magic.

HOW TO BUY AND STORE HONEY

Supermarkets, delis, wholefood shops, health food shops and honey specialists now sell a great range of honeys, across a spectrum of tastes, colours and textures. 'Around the world in 90 pots' (*see page* 174) gives a rundown of the most common kinds, as well as some of the more unusual.

The purest, most natural honey is often that produced by beekeepers who have collected honey from their own hives to sell directly or regionally. They cut the wax top off the honeycombs and spin them, using centrifugal force to extract the honey. It drips down and is left to settle and then may be strained or briefly filtered to remove insect legs and so on, and is then put into pots. Excellent small- or medium-scale commercial honey may go through more filtration but still be treated carefully to retain its flavour and goodness.

Some smaller-scale honey is increasingly labelled as 'raw', 'unfiltered' and 'cold-extracted', which means it has been treated gently and will contain more of its essential nutrients and enzymes. Honey that is left unheated or heated to no higher than around the temperature of the hive (35°C/95°F) is considered to be the best in terms of health properties (*see page* 33) and this is what is meant by 'raw' or unpasteurized honey. When you hold such a pot up to the light you may see that it is slightly hazy with pollen grains.

More mass-produced honeys are heated in bulk and much more finely filtered to clean them up and to make them stay runny for longer. In the most extreme cases, which you shouldn't come across in a decent shop, the honey is diluted and ultra-filtered, sometimes to disguise the fact that the bees have been treated with proscribed chemicals or been imported from an illegal source since this removes the identifying pollen grains. It is then boiled down to the right consistency. This sort of honey tastes a bit like boiled sweets and has none of honey's beautiful characteristics. In the US, it is illegal to call such stuff 'honey'. Some 'honey' illegally uses cheap high-fructose corn syrup to make a more inexpensive product. As with all food, it is best to know that you can trust your sources, in order to get what you've paid for. A good shop or honey brand should take care to source honeys that are what they claim to be.

Decent mass-produced honey is fine to use in any of the recipes in this book, although some recipes that use uncooked honey, for example as a glaze or dressing, will be even better with the high-quality artisanal honeys that are especially fragrant.

The most special honeys of all can be your local ones and any that you buy direct from the producer. One of the best places to buy honey is a farmers' market where you get a chance to talk to the beekeeper. You get a great sense of what has gone into making the honey and, in the case of local honey, the flora around you.

Once you're on the lookout, you can often track down local honey in the most unusual places. Garages, cornershops and village stores may sell a few jars of honey from a local beekeeper, even if just for a couple of weeks a year in the autumn after the honey harvest. Look out, also, for roadside signs advertising honey for sale.

If a friend keeps bees, you may be lucky enough to get hold of one of their pots. Be aware that they may not gather much and want to sell some to offset the costs of their hobby, rather than to give it all away. This is precious stuff produced with much hard labour! Or get in touch with the local beekeeping group and see if they have an event or individuals who sell fragrant honey, fresh from the hive.

Honey festivals, generally held after the autumn harvest or at a time when the beekeepers aren't busy tending their hives, are an interesting way to explore the diversity of honeys and the world of beekeeping.

Some offices, department stores and town halls have honey from bees kept on the roof or in their grounds. Such honey tends to be shared among employees and their friends rather than sold. However, in Saint-Denis, north of Paris, I came across honey from the town hall hives that was given the wonderful name of *miel béton*, or 'concrete honey', and was being sold in the local tourist information centre: a great idea. I was even told which pots were from early summer, later summer and autumn.

Schools that keep bees are another small-scale source of honey. Some head teachers are bringing bees into their grounds and getting the children to tend the insects and put the honey in pots to sell at school events. The honey is a by-product of lessons about nature and pollination rather than a commercial product, but some is available to those connected with the school.

For honey that is more commercially available, there are now a number of extraordinary specialist honey shops, some attached to a beekeeping operation, others run by dedicated honey-lovers with a passion for their subject. Such places gather together not just a wide range of special pots and products such as honeycomb, but also unusual customers and suppliers. You can learn much from stepping into one of these honey-hubs, which also sell a plethora of other bee-related products such as beeswax candles and sweet-scented honey soap. Honey sold by specialists may be available in smaller pots, so you can buy several kinds for the price of a large tub. The quality can be tasted in every spoonful.

If you want to explore the world of honey in greater depth, it is also worth exploring online suppliers. Pots of honey are relatively heavy to send by post, and therefore expensive, but buying online means you can get hold of some of the more unusual kinds from around the world. Specialist websites offer information about particular honeys and tasting notes to guide you through different types. The list of suppliers at the end of the book (*see page* 188) suggests some good places to try.

Many countries sell all their honey locally, rather than exporting it. A number of the honeys in the section 'Around the world in 90 pots' (*see page* 174) are mostly available only in the country where they are produced, although the export market for gourmet honeys is growing.

Be aware that honey is a seasonal product. Look out for it in the autumn, when it has been taken off the hives and is at its most fragrant. It is also worth remembering that some plants only produce nectar in hot summers and some, such as heather, may have a nectar boom-year just one year in seven, and in other years produce nothing at all for the bees. Wherever and whatever kind of honey you find, enjoy it as nature's bounty: it may not be available all year round – or sold at all if the harvest has been bad.

HOW TO STORE HONEY

Honey is its own preservative; it keeps for years. There's a story that honey found in ancient Egyptian pyramids in modern times was still edible. That said, honey is at its very best when fresh, consumed within a year or so of harvesting. This is when the tastes are most complex, vivid and fragrant. But good honey will still be delicious for many years after that.

In terms of texture, many honeys are runny when relatively fresh but granulate naturally over time. If you want a crystallized honey to become runny again, simply put the pot in a bowl of hot water and leave it for 15 minutes or so.

Honey is best stored at room temperature. It tends to crystallize faster at low temperatures. It should also be kept dry, topped with a good lid. Honey is nectar that has been concentrated to around 18–19 per cent water so that it doesn't ferment. Moisture increases the risk of this happening and the honey spoiling. Do not keep your pots in the fridge; this can encourage honey to crystallize, and it may cause condensation and the dilution of the honey, with a risk of fermentation.

HOW TO TASTE HONEY

To appreciate the qualities of different honeys can mean taking just a little more time to taste and enjoy them.

For honey tastings, I take a tub of cut-up straws and tell tasters to dip one end into the honey, like a bee taking nectar from a plant. The honey goes up into the straw and is held there by capillary action. It can then be sucked from the end of the straw, which is either discarded or turned over and the other end used for another honey. This straw 'proboscis' method works best with runny honeys. You can dig your straws into granulated honeys, but for these the best option is to have a pile of clean teaspoons that are used once, then washed after each tasting.

At home, the tip of a teaspoon is the best tool for the job. To compare honeys, you can also dot some of each honey onto a piece of white paper and write its name alongside. Use a spoon or finger to scoop up the honey. You can clear your palate between each honey with a piece of plain bread or a bite of apple. Trained honey-tasters put the honey in a small balloon glass, covered until tasting, and use plastic spoons.

COLOUR

First, look at the honey. Is it light, medium or dark? The colour will vary according to the nectar source or sources.

• LIGHT HONEYS can be a very light gold, defined in the United States as 'water-white', or a slightly stronger yellow colour, like winter sunlight, or pale lemon. Some have a tinge of green and others can be strong and bold, for instance the egg-yolk yellow of sunflower honey.
• MEDIUM HONEYS range from light amber to foxy reds and near-conker. Honey may be brown, but this is a colour with a vast span of shades and hues, be it rust, deep amber or soft gold.
• DARK HONEYS have as many different shades as woods, from oak to mahogany, and be almost black.

A judge at a British honey show is equipped with a pair of grading glasses. These are two small squares that, when held up to the light, show a honey's technical colour category. The strong gold of the lighter grading glass shows the boundary between 'light' and 'medium'. The darker grading glass, the colour of caramelized sugar, shows the difference between 'medium' and 'dark'. The United States Department of Agriculture has seven colour categories for honey: water-white, extra-white, white, extra-light amber, light amber, amber and dark.

These are technical distinctions; at home it's most helpful to put a honey into a rough category and to try to describe the colour. In terms of tasting, a light or medium honey is generally milder in flavour than a dark one, though there are exceptions such as lime honey, which is on the lighter side but has a powerful minty flavour.

Monofloral honeys tend to have a certain colour. But while a honey labelled with the name of a single plant variety should contain mostly the nectar from that plant, it may well have other kinds in the mix, and this will affect the colour of the honey.

Some honeys are creamed (i.e. deliberately granulated) to get a smooth, spreadable texture. This makes the colour lighter than the honey would be in its runny form. Clover and lavender honey, for example, are gold when runny and become creamy white when granulated.

SCENT

Next, smell the honey; this is best done by getting a gorgeous waft straight from the pot immediately after taking off the lid. The scent of a honey, as well as its taste, is notably stronger soon after the harvest, generally in the autumn. Beekeepers talk of the powerful aroma of new honey as they extract it warm from the hive.

Some fresh honeys are so remarkably fragrant that they fill the room with their scent. For several years I was a food writer for the magazine published by Kew Gardens, the world-famous botanical garden and research centre in south-west London. One summer I was given a pot of honey by a staff member who keeps hives there. Lucky bees! A pollen analysis on such a pot would surely be astonishing, given Kew's range of plants from all over the world. On warm days I would occasionally put the pot on my kitchen table with the lid off to fill the air with the scent of flowers.

A strongly scented honey is a clear sign of quality, although some types are inherently more aromatic than others. Interestingly, many honeys do not smell exactly like the plant that provides the nectar, although some do. Orange blossom honey certainly has a floral citrus tang, for example.

If you smell two or three different honeys you can immediately discern the differences between them. The scent of a honey may be resinous, for example heather or chestnut honey; floral; fruity; spicy; mellow; sharp; or medicinal, for example manuka or eucalyptus honey.

TEXTURE

Most fresh high-quality honeys will start off runny and then some crystallize, depending upon the nectar sources and how the honey is stored. Crystallization is not a sign of spoilage or impurity; indeed it may be a sign that this is a very natural honey. Mass-produced honey is fine-filtered to prevent crystallization in order to keep it clear and runny for longer.

Certain honeys also have a gel-like texture. Heather honey, for example, is too gel-like to be spun out of the comb in the normal way and is agitated by little needles, spun carefully or pressed out; it may be beaded with tiny air-bubbles that are trapped in the honey.

The viscosity of honey varies with the nectar source, temperature and moisture content. Honey gathered in colder summers can be more concentrated and viscous because the bees have more time in the hive to concentrate down their honey rather than going out and collecting more nectar. Honey becomes less viscous at higher temperatures.

When scooping, spreading or tasting the honey you can experience its texture. Push your tongue against the roof of your mouth to feel whether it is smooth or has fine or coarse or uneven granulation. Some honeys have a gel-like texture, others are notably smooth and buttery. You can even hear the granules of some honeys, such as sunflower, as you scoop them from the jar and they feel slightly crunchy in your mouth.

TASTE

The scent of honey becomes more defined when you take a taste. To do this in a concentrated way, put a small amount on the tip of your tongue and leave it there for 5–10 seconds or so. You can open your lips slightly and draw in air through your mouth to agitate the flavour molecules as they become warmer and more volatile on your tongue. Then roll the sweet honey juices around your mouth. Flavours take time to develop and the longer you can keep a honey in your mouth, the more you will observe its qualities. This process also happens in almost the same way, of course, when you chew a slice of bread and honey; a more anyday way of tasting honey.

The vocabulary of honey tasting is wide-ranging. Discerning the different tastes is not an exact science, but simply a way to help describe honeys and see how they differ from each other.

You will notice if a honey is mild or strong; simple or complex; gently aromatic or strongly scented or even pungent. Its aromatic components may have berry-fruit or citrus-fruit; a floral nature, with scents of citrus blossom, rose or a more insistent fragrance such as hyacinth. A honey can be herbal or vegetal, with hints of cut grass or green beans. It might be spicy with notes of clove, aniseed, nutmeg or coffee; resinous or woody in a way that is reminiscent of cedar, walnut or pine. The honey may have burnt-sugar touches of toffee, butterscotch, caramel or molasses, or a hint of bitterness or leather.

Some honeys are noticeably more acidic than others and this comes through in the taste. Lime honey, for example, has a strong tang that can be described as minty.

It seems strange to say that you should notice the sweetness of a honey, but honey is composed mainly of two simple sugars, fructose and glucose, in varying proportions. Fructose tastes sweeter than table sugar (sucrose); honeys with a higher proportion of fructose will also tend to remain runny for longer. Acacia honey, for example, is notably sweet and runny because of its high fructose content. On the other hand, chestnut honey is almost savoury, and even slightly bitter, which some people adore and others find repellent.

Sometimes a honey is not at its best. Honey producers test for the HMF factor: this is hydroxymethylfurfural, an organic compound that is produced by the dehydration of sugars. It indicates that the honey has been heated too much or is old; a high HMF count is an indication that the honey won't taste as good as it ought. Honey can pick up scents in its surrounding environment, including smoke used by the beekeeper when harvesting the honey, and it matters that the place where it is kept and processed is clean to prevent funkier savours from tainting the honey.

HONEY TASTING FLIGHTS

An easy way to learn more about honey is to compare different types. To start, try two or three different kinds. A good supermarket, deli or online supplier will have a wide range of honeys to explore. Some online honey specialists sell their own tasting flights and when you go into such a shop they may well have plenty of pots open to sample.

There are several ways to make your own 'tasting flights' of honeys.

LIGHT TO DARK The classic way to taste honey is to go from light-coloured honey to dark. You want to get the delicate floral nuances of light honey before you whack your taste buds with the intense flavours of the dark stuff. Start with a honey such as acacia or orange blossom, then go on to an amber honey such as rosemary and finally stronger or darker ones such as manuka, chestnut or a forest or honeydew honey.

LIGHT, MEDIUM OR DARK It is interesting to taste two or three light or medium or dark honeys against each other to observe the strong differences within each category. A dark eucalyptus honey is very different to Greek mountainside honey, for example, even though they can look similar.

HONEYS WITH SIMILAR NECTAR SOURCES The differences between various flower or herb honeys can be quite noticeable. In the US there are lots of distinctive berry honeys and in Britain you can get both ling heather and bell heather honey. Forest honeys, or honeydew honey, can also be notably different, even if they have the same essential character, depending on whether the woodland is predominantly oak, pine or beech.

HONEYS FROM DIFFERENT COUNTRIES Honey specialists sometimes have interesting imported honeys, such as arbutus honey from Sardinia and Portugal or the butterscotchy honeys from New Zealand flowers such as rewarewa and tawari. When travelling, look for local honeys, especially in places such as farmers' markets. This is a great opportunity to talk to the beekeepers, who will tell you fascinating details about a place and its plants.

HONEY AND HEALTH

Honey has been used for healing for at least 4,000 years. The earliest known reference to its medicinal use is a Sumerian clay tablet dating to around 2000BC that recommends mixing honey with river dust, water and oil, probably to help a skin problem. The ancient Egyptian, Greek, Roman, Arab, Indian and Chinese civilizations all made reference to honey's beneficial properties. To give just two examples, the sixth-century BC philosopher and mathematician Pythagoras is said to have eaten bread and honey for breakfast every day as the basis for a long and healthy life and the philosopher Democritus (c.460–370BC) believed that the secret of his longevity was moistening his skin with oil and his insides with honey.

The association of honey and health persists. There are wild bee nests in the walls of the monolithic rock churches in Ethiopia, sometimes with a gleam of honey quietly dripping from the entrance. The combs are said to produce special healing honey that is used by the priests. In Paris a few years ago I visited La Maison du Miel, a specialist honey shop, and was struck by the number of old people coming in to get a large tub of honey from the drum dispensers in the shop. I was told that they take a spoonful or two every day for good health.

But for most of the twentieth century, honey was largely overlooked, superseded by modern medicine. However, in the 1950s, beekeepers began to explore apitherapy, the use of bee products, including honey, for health. Since then there has been an increased interest in specific types of honey for medicinal purposes and a considerable amount of scientific research into how it works.

ANTIBACTERIAL PROPERTIES

During the First World War honey was commonly used in wound dressings and in recent decades it has been reintroduced into modern medical practice.

All honey has antibacterial properties because its sugars destroy microbes by osmotic force. In addition, honey has the potential to produce the powerfully antimicrobial hydrogen peroxide. This is because, as well as glucose, honey contains glucose oxidase, an enzyme produced by the bees. In the right conditions, this enzyme can break down glucose into hydrogen peroxide. This enzyme is destroyed by heat and so honey that is used for health purposes shouldn't be heated either by the honey packers or by cooking.

The honey with the strongest health pedigree is manuka honey from New Zealand. A Welsh-born biochemist, Dr Peter Molan, began to look into the medicinal properties of honey after he read an editorial in *Archives of Internal Medicine* in 1976 that dismissed it as 'worthless but harmless'. But when Dr Molan looked into the existing scientific literature he found more than 100 studies that suggested it could be actively beneficial.

Furthermore, particular honeys, such as buckwheat, were established in folk medicine as especially healthy.

In New Zealand, manuka honey had been used traditionally for cuts and abrasions. It even smells like antiseptic. Molan and his team tested different honeys by putting them into petri dishes of bacteria to compare how effective they were at killing the bugs. They observed how manuka often acted beyond other honeys and eventually discovered that some manuka honey showed non-peroxide antibacterial activity as well as hydrogen peroxide activity. Manuka honey that has both activities is especially effective. The researchers developed a UMF (unique manuka factor) rating; the higher the better, and there are also other grading systems. Such research means that manuka honey is now certified for use in wound dressings in the UK, the US and elsewhere. Manuka honey is also used by veterinarians on animals. There is, however, some fake manuka sold; it can be a victim of its own marketing success, so buy from a trustworthy source.

Other honeys now also have laboratory evidence behind them. For example, jarrah honey, produced from a tall and beautiful eucalyptus tree in Western Australia, has been tested for its medicinal properties and other proven 'healthy honeys' are likely to emerge. As well as being useful for wound dressings and external use, honey may kill some of the bacteria that cause stomach problems and irritable bowel syndrome. While this is far from being scientifically proven, we are, to some extent, our own laboratories as far as day-to-day health is concerned. Try it and see if it works for you.

TRADITIONAL MEDICINE

Kew Gardens in London has a research team looking at the traditional uses of plants around the world. Professor Monique Simmonds told me how honey is used alongside plants in traditional medicines. The sugars in the honey help burst the cell walls of the plants and in this way may help their bioactive properties to be more effective. A spoonful of honey really does help the medicine go down, and not just because it makes bitter compounds taste better.

Sweets are now seen as an indulgence, but their history reveals close links between the apothecary and the confectioner, with honey a key ingredient. Lozenges and cordials, some using honey, were meant to make you feel better. Honey is stirred into herbal teas; its sweetness brings out the flavour of the leaves as well as making their health aspects more bioavailable.

SWEETER THAN SUGAR?

These days sugar is regarded as a nutritional no-no because of its 'empty calories' that deliver energy but no other nutritional benefits. Too much sugar may disrupt your hormone system and many people believe that soft drinks and processed carbohydrates, especially from the sugar in cakes and biscuits, are behind the current epidemics of obesity and diabetes.

Yes, honey is sugar – natural glucose and fructose, but sugar nonetheless. But it also contains small amounts of vitamins, minerals, phytochemicals and beneficial enzymes. It has a stronger taste than refined table sugars (sucrose) and a little can go a long way. Many of the recipes in this book use just a spoonful or two and treat precious honey as a seasoning. When people object that honey is 'just sugar', it is worth considering how much of it is used, as well as its other nutritional components. A little honey, added to a dish to make it even more delicious, is different to gulping down large glasses of sugary pop.

The rate at which different foods raise blood sugar levels is measured on a scale called the glycaemic index (GI). Pure glucose is counted as 100 and anything above 70 is considered a high-GI food; the lower the GI the better. Honey's GI is typically 58, around the same as table sugar, but it can vary between 32 and 85, depending on the nectar source. Fructose-high honeys such as acacia or black locust and yellow box have the lowest GI, in relative terms.

BEE AWARE

Other products of the hive are used in apitherapy. Propolis is a resin exuded by plants, which bees collect and use to plug gaps in their hive. Beekeepers with a tickly throat take a scrape of propolis from the hive as an antiseptic and it is sold as a tincture and in other forms, such as lozenges.

Pollen is the protein-rich substance that bees gather from plants to feed themselves and nourish their larvae. The little balls of pollen that are collected and kneaded together by the bees are sold as a health food. I like it served on porridge and in smoothies and sometimes use it as a garnish on soup, both for colour and nutrition. Some people believe that eating pollen or honey rich in pollen protects against hay fever, especially when from the local area. The second-grade honey that comes from the pollen-rich cappings taken off the combs is thought to be especially helpful and is sometimes available from beekeepers.

Without overplaying honey and other hive products as some sort of elixir of life, they do have properties that have become overlooked but were much more valued in the past. Furthermore, a respect for honey is part of a gentler and more open approach to the world around us. Watching bees on flowers, valuing what they give to the kitchen and being aware of how they fit into the intricate ways of nature are part of slowing down and finding a sense of harmony with our surroundings; Bee-ing not doing can contribute to our well-being and health.

HONEY AND THE NATURAL WORLD

Honey brings creatures, plants and humans together. Bees are thought to fly 55,000 miles, gathering nectar from some two million flowers, to make a single 450g/1lb pot of honey. Each of these bees, busy as they are, makes about one-twelfth of a teaspoonful of honey in a lifetime.

Honeybees gather nectar from within a few miles of their nest or hive. The Ancient Laws of Ireland said that a bee flew as far as the sound of a church bell, or a cock's crow. Science now shows they generally fly within a half or one mile, and up to three. Good honey from a single location is a quintessential local food; a unique taste of time and place.

Just as your garden and the countryside vary with the time of year, honey gathered from the same place in different seasons is strikingly different. Early summer honey, made when the apple trees and blackberry bushes are blossoming, is like a fantasy honey-pie. Late summer honey and autumn honey each have their own distinctive flavours. Honeys from tropical countries are as varied as the tropical flora, sometimes with distinctive combinations. The queen bee of honey history, Eva Crane (1912–2007), writes of honeys that mix orange and coffee blossoms. Honeys from the Alpine and Greek mountainside or Welsh countryside will contain wonderful mixtures of wild flowers.

Whenever you talk to a beekeeper you hear about the plants of a place and what's going on in nature. Much of this brings a delightful sense of how different floral scents and savours end up on your toast. Other sources of honey are odd yet engaging. One beekeeper told me that his bees mass on privet hedges that have been left untrimmed and allowed to flower. At first the honey tastes awful, then it mellows and becomes palatable. One year, a chocolate factory sold unwanted coloured fondant eggs to a local pig farmer; his bees produced pink honey. American bees have been known to swoop into vats of sweet drinks, resulting in a bright pink cherry pop honey.

Beekeepers are not the only people who are fascinated by these insects. It's been said that when something is overlooked it is the poets, the naturalists and children who still see it. How often have I seen a child crouched down looking at a bee! Once more aware, you find yourself watching these insects in the air as they fly and as they concentrate on gathering nectar and pollen, undisturbed by your attention. See bees; see life.

It is a revelation to look at the natural world from a bee's-eye view. Marjoram gives out the sweetest nectar, generally at about midday, and I see the bees on my herb patch before I prepare my own lunch. Ivy attracts bees with a strong burst of sugar late in the year. Look at the small yellow-green crown of flowers on ivy in September and you can sometimes see them drip with nectar. This will help keep the bees alive when they cluster around their queen during the cold days until the air warms and nectar flows again in spring.

Different types of bees feed on different flowers; the long-tongued bumblebees are able to drink from the nectaries at the base of deep flowers such as foxgloves, which honeybees can't reach. There are around 270 bee species in Britain and 25,000 throughout the world. The honeybees are the ones that work together to make and store honey but all bees help to keep the planet alive.

Bees and flowers co-evolved in the Cretaceous era and their lives are much interconnected. Look at bees' hind legs and you can sometimes see little balls of pollen. This highly nutritious substance is the bees' protein source. It sticks to their hairy bodies and is swept down to the 'pollen baskets' on their back legs. Pollen's primary role is the fertilization of plants. As bees fly from one flower to another, the pollen brushes onto the plant's female parts and fertilization takes place.

Four-fifths of the world's plants rely on pollination to reproduce. Many other insects, such as ants and moths, as well as birds and other animals, pollinate plants but bees, and specifically honeybees, are the most manageable creatures dedicated to this task. According to the Food and Agriculture Organization of the United Nations, at least one third of the world's food crops depends upon pollination. The economic value of bees worldwide is estimated at US$217 billion because of all the food they help produce. Without bees we would have staples such as wheat and rice but would lack fruits, nuts, edible seeds and their oils, leafy greens and such common vegetables as onions. A plate would look very bare without the foods that rely on insects for pollination. These creatures buzzing through the air are the stitches that hold the fabric of the living world together. They animate our planet and keep it strong.

PLANTING FOR BEES

You don't need to become a beekeeper to look after bees. Look at your garden or plant pots from a bee's-eye view and you can learn how to provide them with nectar and pollen.

Colour is very important to bees. They can't see red but at the other end of the light spectrum they can see the ultraviolet that is invisible to human eyes. A field poppy, for example, is bright red but the bee is drawn to the ultraviolet at its centre and the dark blue pollen. Generally, yellow, white, blue and, in particular, purple seem to be the bees' favourite colours. Observe the flowers they go to in your local parks, gardens and countryside and you'll begin to get an idea of what to plant.

Flowers in nature evolved to attract bees. Some modern cultivated plants are designed to please the human eye, but double-flowered varieties, for example, mean the bees can't get to the nectar. Go for varieties that aren't just ornamental.

Plant so that the bees have food throughout the year. They need food from as early as January and as late as a mild November. They are especially hungry in the spring when they first emerge from their nest or hive, so early-flowering plants such as crocus, snowdrop and early-flowering viburnum help the bees on bright spring days. What is known as the 'June gap' is a surprising patch of hunger for the bees, when the food from some trees and spring flowers has gone but the main summer blossoming hasn't started. Herbs such as thyme, sage, oregano and mint are useful here, as well as borage and phacelia, a plant with purple-blue flowers that always throngs with bees. At the end of the season, Michaelmas daisy, aster, golden rod and ivy will feed the bees on warm days late in the year.

Many books go into great detail about various plants to feed bees in different climates. *Plants for Bees*, published by the International Bee Research Association (IBRA), is the best one for the British Isles, and the Royal Horticultural Society's website (www.rhs.org.uk) includes a list of plants to attract bees into the garden. Sarah Wyndham Lewis gives many useful ideas in *Planting for Honeybees*.

Wildflower patches matter; a bumblebee is only 40 minutes away from starvation because of all the energy it uses to keep flying and honeybees like this sort of forage, too. Simple changes such as not mowing the lawn until after any clovers have finished flowering are ways that gardeners can help bees. Better still if you can manage part of your lawn in a larger garden as a wildflower meadow, or encourage the people who manage your local parks or countryside to foster wildflowers in this way.

In cities, trees are an important source of food for bees, as well as protecting us from pollution and providing habitats for wildlife. Urban development all too often happens without consideration for the plants and places that provide food and shelter and water for creatures. Yet this perspective is crucial if we are to keep the world and our surroundings healthy.

BEES: THEIR PROBLEMS ARE OUR PROBLEMS

The way we see bees and honey reflects society's concerns. This has always been the case, altering as society changes. The Romans admired the bees' bravery in being prepared to die to protect the colony, like a legionnaire in battle. Seventeenth-century rationalists admired the way in which the different parts of the colony worked together. The Victorians believed bees to be pious creatures that would sting anyone who swore near the hive. After the Second World War and the experience of totalitarian regimes, bees were sometimes seen in a more sinister light owing to the way they act as one single ruthless force.

Today, our concerns about bees mirror fears about the state of the environment. Bees do die over winter; beekeepers expect to lose 10–15 per cent of their bees as a matter of course. Incidents of large-scale deaths and disappearances have always been known, but over the past ten years or so bee losses have escalated. In the United States, heavy winter losses, starting in 2007–9, led to the use of the term colony collapse disorder (CCD), when beekeepers lost on average 33 per cent and in some cases up to 90 per cent of their bees. The main symptom of CCD is very few adult honeybees present in the hive and no dead honeybee bodies, but with a live queen – in other words the bees seem to have simply disappeared. Not surprisingly, this resulted in some apocalyptic headlines. Before the advent of CCD, French beekeepers were up in arms about widespread, serious bee losses that they blamed on pesticides.

No single reason has been found for this large-scale problem and there are probably a number of causes.

Honeybees are not native to the New World, but are increasingly important for the pollination of food crops. Almonds, for example, are dependent on honeybees for pollination, and 80 per cent of the world's almonds are grown in California. To meet global demand, California's almond growers need around 1.5 million honeybee colonies and beekeepers drive their bees here from all over the country at the start of the beekeeping year. Beekeepers have long carted large numbers of hives around the country to pollinate crops of many kinds. Some researchers link the stress of being transported to the increase in CCD. Others suggest that the huge concentration of hives is resulting in the rapid spread of diseases and parasites.

Stress of many sorts is thought to affect today's bees. In recent decades there has been an increasing problem with the varroa mite, a parasite that harms both the bees and their brood and spreads viruses throughout the colony. This has weakened bees around the world, especially in the northern hemisphere (African honeybees seem to be resistant and Australia has avoided varroa as yet).

Habitat loss is the main cause of concern. In the past 50 years, Britain has lost 90 per cent of its wildflower meadows, as land is more intensively farmed or built over. Miles of hedges have been grubbed up and thousands of trees felled. Scientists and some farmers are now working out how to produce food in a more wildlife-friendly way. But from a bees'-eye view, large parts of the countryside are bare of variety and food and this can leave them ill-nourished. Different bees pollinate different plants, and it's important that bees of all kinds are able to feed, in order to protect biodiversity. Climate change also affects bees because of their forage and the way that they are greatly affected by weather and temperature.

And then, of course, pesticides and herbicides are used specifically to attack insects and the plants that are their food sources. A group of insecticides, developed from a synthetic form of nicotine, called neonicotinoids (or 'neonics'), have been sprayed onto plants, farmland and gardens and also used to coat seeds and to soak bulbs, so they are 'systemic' and go throughout the plant, including the nectar. The makers of these pesticides say they have been tested to contain chemicals at a level that is 'sub-lethal' to honeybees. Leaving aside their effects on other creatures, scientists are discovering that these products seem to be a problem when combined with other elements such as bee disease and even the miticides and other treatments used by beekeepers to combat disease in the hive.

Neonicotinoids are now to be banned from outdoor use by the EU due to the mounting evidence of their damage to pollinators, and the UK government has changed its policy to support this.

As consumers, cooks and honey lovers, we can follow the debates and perhaps choose to buy food and drink that are grown in a way that doesn't put bees at risk, such as organic produce. As gardeners, we can question nurseries and try to avoid pesticides and herbicides that may put bees in danger, as well as growing plants that provide nectar and pollen to feed bees. As citizens, we can ask questions and push forward the principle that nature is not a commercial commodity but the present and future of the world itself.

Honey and bees can show us the interconnections and vulnerabilities of the world around us, as well as celebrating the extraordinary power of nature.

BREAKFAST AND BRUNCH

RICOTTA HOTCAKES WITH
HONEY-GLAZED BACON

Pancakes and their tubbier cousins, hotcakes, are the perfect brunch dish: filling, comforting and a vehicle for both sweet and savoury toppings. Drizzle honey over the hotcakes with some melting butter and then add some salty-sweet bacon that has been caramelized with a little honey. Alternatively, instead of bacon, honey and berries are great with these hotcakes.

In Sicily, they eat very fresh ricotta with a drizzle of honey for breakfast. The ricotta you get in tubs isn't quite up to this sort of sublime simplicity, but it is excellent as an ingredient in many dishes. It gives these hotcakes a creaminess and makes them more robust, in the lightest possible way.

SERVES 5–6 (ABOUT 15 HOTCAKES)

250G/9OZ RICOTTA CHEESE

150ML/5FL OZ FULL-FAT MILK

125G/4½OZ PLAIN FLOUR

LARGE PINCH OF SEA SALT

2 LARGE EGGS

A COUPLE OF PATS OF UNSALTED BUTTER, PLUS EXTRA TO SERVE

SUNFLOWER OR VEGETABLE OIL

8–12 RASHERS OF SMOKED STREAKY BACON

1½ TBSP RUNNY HONEY, PLUS EXTRA TO DRIZZLE

In a large bowl, mix together the ricotta and milk. Stir in the flour and salt. Separate the eggs, putting the whites into a clean bowl and stirring the yolks into the ricotta mixture. Whisk the egg whites until stiff, then gently fold them into the ricotta, keeping as much air in as possible. You can do this up to an hour in advance of cooking, but keep the mixture, covered, in the fridge.

Melt a pat of butter with a little oil in a frying pan over a medium–high heat. Dollop large tablespoonfuls of the batter into the pan. It will make hotcakes 6–7cm/2½in diameter and 1.5cm/⅝in thick. You don't need to spread them out; leave them as tubby little clouds that find their own form. Cook the hotcakes for 3–4 minutes, until small holes appear on the top. Turn them over carefully, using a spatula. Cook on the other side for 2 minutes, until patched with brown. Continue to cook the rest of the batter, keeping the cooked hotcakes warm, and adding more butter and oil as necessary.

Meanwhile, in another frying pan over a medium heat, cook the bacon on both sides, until brown. Spoon over the honey and quickly turn the bacon over. Cook for 30 seconds to a minute, until the bacon is browned. Remove from the pan. Serve two or three hotcakes per person, with a little more butter, the bacon and a drizzle of honey on top.

AMLOU
(MOROCCAN NUT AND HONEY BUTTER)

This fragrant spread, a combination of honey, toasted almonds and nut oil, is a treat from the Moroccan breakfast table. In Morocco the nuts are pulverized using a mortar and pestle to get a very fine powder, but this is hard to do and so here is a slightly chunkier version, akin to crunchy peanut butter. Easy to make, healthy and nourishing, Amlou is typically served in a small bowl alongside layered pancakes that are shiny with matured butter, smen, or with semolina-crusted bread.

The spiky, gnarled argan tree grows in the south-west of Morocco. Locals have long prized the oil made from its nuts and use it both cosmetically and in cooking – to anoint tagines or to make this unusual nut butter. The argan is one of the toughest of all nuts to crack. Teams of women in Morocco still do this by hand using rocks: no wonder argan oil is expensive. But it is worth seeking out. The aromatic taste takes you straight to the souks and the bright, dry countryside of Morocco. Argan oil is now considered to be a health food and is sold around the world. You can find it in specialist shops and online (see Suppliers, page 188). Make sure you get the culinary grade and not the stuff that's meant for hair and skin.

All nut oils deteriorate over time and need to be used within six months or so. Amlou is an excellent way to use up argan oil. In the West, I've also seen it used in ice cream and pannacotta, both of which could combine argan oil successfully with honey.

MAKES A SMALL BOWLFUL OR A MEDIUM JARFUL

125G/4½OZ BLANCHED ALMONDS

LARGE PINCH OF SEA SALT

4 TBSP ARGAN OIL

40G/1½OZ HONEY

Preheat the oven to 160°C/325°F/Gas 3. Spread the almonds on a baking sheet and toast in the oven for 15 minutes, or until lightly brown, turning the baking sheet around halfway through to ensure the nuts cook evenly and taking care they do not burn. Leave to cool.

In a food processor – ideally a small one – grind the roasted almonds. Pulse the machine to grind the nuts as finely as you can, but don't over-process them.

Tip the ground almonds into a small serving bowl. Add the salt and then the oil, stirring to get a slightly liquid paste. Add the honey, mix well and taste to see if it needs more salt. Cover the bowl – or store in a jar with a lid – and keep in a cool place, but not the fridge.

Serve with pancakes or Moroccan bread for breakfast or brunch, or with any kind of flatbread, lightly browned and warmed through on a griddle.

HONEY CINNAMON BUNS

The scent of these buns calls you to the breakfast table in the most tempting way. These are great made the day before eating, but if you want that just-baked freshness, you can make the dough the night before and leave it to rise in the fridge overnight in a large bowl, covered loosely with clingfilm. Take the dough out early the next morning so it can warm up before you get it going again.

MAKES 12

400G/14OZ STRONG WHITE BREAD FLOUR,
PLUS EXTRA FOR DUSTING

½ TSP FINE SEA SALT

6G EASY-BLEND DRIED YEAST (MOST OF
A 7G SACHET)

10G/¼OZ UNSALTED BUTTER, CUBED

ABOUT 250ML/9FL OZ WARM WATER

A LITTLE VEGETABLE OIL FOR GREASING

HONEY-CINNAMON FILLING

100G/3½OZ CURRANTS

4 TBSP HONEY (A MEDIUM OR DARK HONEY
SUCH AS HEATHER OR THYME WORKS WELL)

3 TBSP WATER

50G/1¾OZ UNSALTED BUTTER, AT ROOM
TEMPERATURE

¾ TSP GROUND CINNAMON

GLAZE

1 EGG

2 TBSP HONEY

For the filling, put the currants, 1 tbsp of the honey and all the water in a small pan. Bring to the boil, then turn the heat to low and leave, covered, to allow the honey syrup to plump up and flavour the currants.

Put the flour, salt and yeast in a bowl. Rub in the butter using your fingertips and add 190ml/just under 7fl oz of the warm water. Bring the ingredients together and add as much of the rest of the water as you need to make a dough that comes away from the sides of the bowl. Knead for 10–15 minutes (8–10 minutes in an electric mixer with a dough hook), until the mixture becomes smooth and less sticky.

Put in a large, lightly oiled bowl and put the bowl inside a large plastic bag that won't touch the top of the risen dough, for 2 hours, or until roughly doubled in size. You can make the rise happen faster in a warmish place, but a slower rise will give more flavour to the buns.

While the dough is rising, prepare the filling: mix together the butter, cinnamon and remaining 3 tbsp honey. Lightly dust the work surface with flour and tip the dough onto it. Knock out the air with your fist and by folding the dough over onto itself several times. Roll out the dough into an oblong about 25cm/10in wide by 35–40cm/14–16in long. Put spoonfuls of the honey and cinnamon butter all over the dough, then spread all over the surface with the back of the spoon. Scatter over the plumped-up currants.

Line a baking sheet with baking parchment. Roll up the dough along the long side. Use a large sharp knife or baker's scraper to cut into four pieces. Cut each of these pieces into three equal slices, so you have 12 pieces of dough, and place on the baking sheet with 3–4cm/1½in between each one. Put the baking sheet in a large plastic bag – loose enough not to touch the risen dough – and leave to rise in a warmish place for 30 minutes. Towards the end of this time, preheat the oven to 220°C/425°F/Gas 7.

Meanwhile, make the glaze: beat together the egg and honey. Brush this over the risen buns, taking care not to knock out the air. Carefully poke any currants down into the gaps in the dough (exposed currants will burn in the oven). Bake for 15–20 minutes. The honey glaze will make the buns very dark, so turn the baking sheet around in the oven once or twice to make sure they cook evenly and don't burn.

Remove from the oven and leave to cool on the baking sheet. Remove any burnt currants from the surface of the buns, then tear the buns apart. Serve either warm or cold.

HONEY GRANOLA

Granola is very useful to have on hand as a breakfast cereal that's full of energy to get you going in the morning. I like a couple of spoonfuls of this on top of yogurt and fresh fruit, or you can soften the grains slightly with milk and add some chopped fresh banana and berries. This sweet and tasty honeyed mixture is also excellent to add crunchy texture and extra flavour to poached fruit or a fruit salad as a pudding.

Making a batch of granola is a great way to use up the bags of dried fruit, nuts and seeds that seem to accumulate in the kitchen cupboard and I've written the recipe so as to leave the choice of nuts, fruit and seeds up to you. I tend to go for two types of each. My standard mixture is pistachios and almonds, apricots and cranberries, and sesame and sunflower seeds. Sometimes I go tropical and include pecans and dried mangos in the mix and add dried coconut shavings with the fruit.

MAKES A 500ML/18FL OZ JARFUL

1 TBSP SUNFLOWER OIL, PLUS EXTRA FOR GREASING

2 TBSP HONEY

1 TSP VANILLA EXTRACT

100G/3½OZ ROLLED OATS (JUMBO OATS RATHER THAN QUICK-COOK PORRIDGE OATS)

50G/1¾OZ NUTS, ROUGHLY CHOPPED (SMALL NUTS SUCH AS HAZELNUTS OR PISTACHIOS CAN BE LEFT WHOLE)

30G/1OZ SEEDS, SUCH AS SESAME, SUNFLOWER, LINSEED, OR MIXED SEEDS

50G/1¾OZ DRIED FRUIT OR BERRIES, OR LARGER DRIED FRUIT CHOPPED INTO PIECES ABOUT THE SIZE OF A HAZELNUT

Preheat the oven to 140°C/275°F/Gas 1. Lightly oil a baking sheet.

Put the oil and honey in a small pan, measuring out the oil first and then using the same spoon to measure out the honey so it slips easily off the spoon. Heat gently for 30 seconds or so, stirring, to loosen and combine the two. Stir in the vanilla extract and take off the heat.

Put the oats, nuts and seeds into a big bowl. Pour over the honey mixture and use a wooden spoon or a hand to mix the honey thoroughly and evenly into the dry ingredients. Spread the honeyed mixture over the baking sheet. Toast in the oven for 15–20 minutes, or until lightly brown, stirring two or three times during this period so that the mixture cooks evenly.

Stir in the dried fruit and leave the mixture on the baking sheet to cool, stirring it around a couple of times so it doesn't stick together too much. This keeps well in an airtight container for a couple of months.

MUHAMMAD ALI SMOOTHIE

Muhammad Ali, world heavyweight champion boxer, didn't just 'float like a butterfly and sting like a bee', he ate like a bee, too, regarding pollen as a powerful health food. This smoothie is named in his honour.

Pollen grains contain a plant's male reproductive cells and provide the bees' prime source of protein. Bees emerge from plants dusted with pollen that is caught in their hairy bodies and they sweep this down to little indentations, called pollen sacks, on their back legs. You can spot these colourful pollen loads on bees as they fly between plants. It is collected by the beekeeper by putting a small comb in front of the hive entrance from time to time so the pollen is knocked off the bees' legs.

High in protein and other nutrients, pollen is sold as a health food for humans. It is great scattered over breakfast porridge or other cereals and looks particularly pretty on top of this refreshing bright pink smoothie. Once you get into pollen, you find many different ways of using it throughout the day. Don't eat too much, however: half a teaspoon at a time is enough. Good as it is to have energy and strength from the get-go, those who eat a lot of pollen say they feel a touch too 'buzzy'.

SERVES 1

50G/1¾OZ FROZEN RASPBERRIES

4 TBSP PLAIN YOGURT

4 TBSP APPLE JUICE, OR MORE, TO TASTE

1 TSP FRESH LIME JUICE

1–2 TSP HONEY

½ TSP POLLEN

Put the raspberries, yogurt, 4 tbsp apple juice, lime juice and honey in a blender with half the pollen.

Whizz up the mixture until the raspberries are blended with the other ingredients. Check the consistency and add more apple juice if you want a more liquid smoothie. Taste the drink and add more lime juice or honey, if necessary.

Serve the smoothie in a glass and scatter the remaining pollen on top. Knockout!

AMBROSIAL APRICOTS WITH THYME

There's an affinity between herbs and honey, as anyone knows after watching bees buzz between flowers in a herb patch.

The fragrance of the apricots, honey and thyme combine to give you the best possible start to the day. Sometimes all you want is just a few spoonfuls of this dish, but sometimes it is nice to make it more substantial and creamy with a dollop of thick yogurt.

If you want to serve this as a dessert, then a dash of gin, added at the end, does wonders for apricots, and the botanicals in the gin enhance the herbal tang of the thyme.

SERVES 4

500G/1LB 2OZ RIPE (OR NEARLY RIPE) APRICOTS

300ML/10FL OZ WATER

4 TBSP HONEY (A MEDIUM OR DARK ONE WORKS WELL AND GREEK THYME HONEY IS IDEAL)

4 SPRIGS OF THYME (WITH FLOWERS IF HOME-GROWN)

GREEK-STYLE YOGURT (OPTIONAL), TO SERVE

Cut the apricots in half, if large. You can remove the stones, but I tend to leave them in, both because they are easier to remove from the cooked fruit and to let people enjoy nibbling the flesh off the stone.

Put the apricots in a medium–large pan, ideally in a single layer. Pour over the water and drizzle over the honey. Tuck the sprigs of thyme among the apricots. If you have the thyme flowers, remove these to use as a garnish.

Put a lid on the pan and bring slowly just to the boil. By the time it reaches boiling point, the apricots should have softened slightly; turn off the heat and carefully turn them over. Cover and leave for 10 minutes so the lower half cooks in the residual heat. You don't want to overcook the fruit.

Remove the apricots with a slotted spoon and put them in a serving bowl. Turn up the heat under the pan and bring to the boil. Boil for 5 minutes, or until the juices are slightly syrupy. I prefer to keep the mixture quite light rather than over-reduced. Leave the syrup to cool and infuse with the thyme.

Pour the syrup over the apricots. Remove the thyme sprigs. Some of the leaves may have fallen off but if not, pull some off to scatter over the dish, if you like. If you have thyme flowers, scatter them over now.

These ambrosial honeyed apricots are best at room temperature but fine served chilled from the fridge; add a good spoonful of yogurt, if you like.

LUNCH AND SUPPER

HONEY AND TAMARIND RIBS

Honey combines with tangy tamarind to make this finger-licking rib glaze. The sweet stickiness and crusty surface contrast happily with the soft, melting fat and toothsome flesh of the ribs. The ribs are first simmered, gaining flavour from a few aromatic ingredients in the cooking water: star anise, cinnamon and/or a couple of bay leaves; I generally go for all three, but my favourite is the star anise. The cooked meat absorbs the honey glaze beautifully. If you have time, marinate the meat before cooking. Tenderness and taste are what you're after in ribs.

SERVES 4

1.4KG/3LB 3OZ PORK RIBS, EITHER SINGLE RIBS OR 4 SMALL RACKS

1 TBSP SALT

ONE OR MORE OF THE FOLLOWING: 1 STAR ANISE, ½ STICK CINNAMON, 2 BAY LEAVES

TAMARIND GLAZE

1½ TBSP TAMARIND PASTE

3 TBSP HONEY (MEDIUM OR DARK AND STRONG ARE BEST)

1 TBSP SOFT BROWN SUGAR

2 GARLIC CLOVES, CRUSHED

1 TBSP OLIVE OR VEGETABLE OIL

1 TBSP SOY SAUCE

2 TBSP WATER

Put the ribs in a large pan and add the salt and the star anise, cinnamon and/or bay leaves. Cover with cold water. Bring to the boil, turn down the heat and simmer for 1 hour. Meanwhile, make the glaze by mixing all the ingredients together in a large sealable plastic freezer bag (or in a large shallow dish).

Drain the ribs, reserving the liquid to use as stock in another dish. Discard the star anise, cinnamon stick and bay leaves. Add the ribs to the glaze, turning them around to ensure they are evenly covered. This is easiest to do in a bag, as you can massage them all over.

You can cook the ribs straight away, or leave them to marinate in the glaze. Seal the bag or cover the dish and, once the meat is cool, leave in the fridge for up to 24 hours. Turn the meat over every so often.

Preheat the grill to high, or light your barbecue and wait until the flames have died down and the ash-grey charcoal pulses with heat. Line a baking sheet with foil folded up at the edges to contain the marinade, then lay out the ribs and grill them not too close to the heat, or barbecue them, basting often, until the meat is hot all the way through and the glaze is thick and sticky (10–15 minutes), turning them every few minutes so they cook evenly.

ROMAN HONEY-BAKED HAM

The ancient Romans used plenty of honey in cooking and honey is still used widely in food in Rome. A recipe from Apicius, the collection of recipes thought to be from around the end of the 4th century AD, recommends spreading honey over a cooked ham and then covering it with a paste of flour and water before baking. I've tried this method and it works very well, retaining the freshness of the honey. This recipe uses the same idea but with a breadcrumb crust. The original Roman recipe includes dried figs and bay leaves in the cooking stock for the ham and they both add a lovely flavour and fragrance to the meat.

SERVES 5–6

750G/1LB 10OZ GAMMON (SMOKED OR UNSMOKED)

3 BAY LEAVES

STALKS FROM A BUNCH OF PARSLEY

8 DRIED FIGS (OPTIONAL)

400ML/14FL OZ APPLE JUICE

ABOUT 1 LITRE/1¾ PINTS WATER

HONEY-BREADCRUMB COATING

70G/2½OZ COARSE WHITE BREADCRUMBS

1½ TBSP DIJON MUSTARD

1 TBSP OLIVE OR SUNFLOWER OIL

2 TBSP HONEY

FRESHLY GROUND BLACK PEPPER

Put the gammon in a pan and add the bay leaves, parsley stalks, figs and apple juice, topping up with water to nearly cover the gammon. Bring to the boil, then turn down the heat, put a lid on the pan and simmer for 1¼ hours. Remove the gammon from the pan and leave to cool slightly. Strain off the cooking liquor; you can cool this stock and keep it, covered, in the fridge to make a tasty soup. Preheat the oven to 190°C/375°F/Gas 5.

Put the breadcrumbs in a bowl and add the mustard, oil, 1 tbsp honey and a good grinding of black pepper. You won't need any salt, as the gammon is salty enough.

Cut the rind off the gammon, leaving a good layer of fat. Use a sharp knife to cut a criss-cross pattern into the fat. Spread 1 tbsp honey into the fat, then cover the top of the gammon with the breadcrumb mixture, pressing it down well; the honey will help it stick.

Bake in the oven for about 10 minutes, until the crumbs are lightly brown and crisp. Remove from the oven and leave for at least 10 minutes before carving into thin slices. Serve hot, warm or cold, making sure each person gets a good piece of crumb coating alongside their meat.

LAMB AND HONEYED APRICOT TAGINE

Fruit in tagines is often cooked separately with aromatic flavourings and added at the end to give the sweet-and-savoury balance that is so characteristic of Moroccan cooking. Here, apricots in a honeyed syrup with juniper and orange blossom water add a final layer of fragrance. There isn't much honey in this tagine, but it is a crucial element in a dish that is simple to make but as sophisticated as the sort of food you get in Moroccan homes.

I've been lucky enough to watch some of Morocco's superb home cooks at work and pick up some tagine tips. First, don't use too much water: the ingredients mostly cook in their own juices. Grating rather than chopping the onion gives a more melded sauce, which clings to the meat rather than swamping it. The spicing of Moroccan food has fragrance rather than heat. This recipe uses the basic but subtle Berber blend of ginger, turmeric and pepper.

SERVES 4

1 LARGE ONION, COARSELY GRATED

2 GARLIC CLOVES, FINELY CHOPPED

1 TSP GROUND GINGER

1 TSP GROUND TURMERIC

1 TSP FRESHLY GROUND BLACK PEPPER

125ML/4FL OZ WATER

900G/2LB HALF LAMB SHOULDER

½–1 TSP SEA SALT

5 CARROTS

2 TBSP CHOPPED FLAT-LEAF PARSLEY

1 TBSP CHOPPED FRESH CORIANDER

HONEYED APRICOTS

12 DRIED APRICOTS

1 TBSP HONEY

2 TBSP WATER

1 TSP ORANGE BLOSSOM WATER

5 JUNIPER BERRIES, LIGHTLY CRUSHED (OPTIONAL)

Preheat the oven to 180°C/350°F/Gas 4. Put the onion into a tagine or casserole with the garlic, ginger, turmeric, pepper and water and mix to a paste. You can leave the lamb shoulder whole, but I prefer to cut four large chunks (about 9cm/3½in long) off it, leaving the bone with the rest of the meat attached. Rub the paste all over the lamb. If you have time you can leave this to marinate for an hour or so, or longer in the fridge, covered.

Season the meat with salt. Scrunch up a piece of greaseproof paper slightly bigger than the diameter of the pan, wet it under the tap and place it over the meat. Cover the tagine with a lid and put in the oven for 1 hour.

Meanwhile, put the apricots in a small pan, add water to cover, bring to the boil and turn off the heat. Leave to soak.

Cut the carrots into large chunks about 8cm/3¼in long, on the diagonal. Add the carrots to the tagine after it has cooked for 1 hour. Re-cover with the greaseproof paper and cook for another 1¼–2 hours, or until the meat falls away from the bone.

While the meat is cooking, drain the apricots and then put them back in their pan with the honey, water, orange blossom water and juniper, if using. Bring to the boil and boil for 5 minutes.

When the lamb is tender, stir in the honeyed apricots and scatter over the chopped herbs. Serve traditionally with couscous, or with rice, or with potatoes baked in the oven alongside the meat.

THYME-GLAZED ROAST LAMB WITH HONEY AND CIDER GRAVY

The sweetness of roast lamb is enhanced by a honey glaze scattered with thyme and a sharp-sweet gravy made of cider and honey. This dish is inspired by the association of Wales with bees and honey, as well as lamb. The International Bee Research Association (IBRA), founded in 1949 by Dr Eva Crane, is based in Cardiff and IBRA's remarkable bee- and honey-related library is now housed at the National Library of Wales in Aberystwyth. It consists of some 4,000 books and 50,000 academic papers and periodicals relating to insects and honey from around the world.

SERVES 6, WITH LEFTOVERS
6–8 MEDIUM–LARGE FLOURY POTATOES, PEELED

SEA SALT AND FRESHLY GROUND BLACK PEPPER

2KG/4LB 8OZ LEG OF LAMB

3 TBSP OLIVE OIL

2 BUSHY SPRIGS OF THYME

1 TSP PLAIN FLOUR

1 TBSP HONEY (THYME HONEY IS PERFECT)

HONEY AND CIDER GRAVY
½ TBSP PLAIN FLOUR

250ML/9FL OZ DRY CIDER

200ML/7FL OZ STOCK (ANY KIND, BUT BEWARE OF OVERSALTINESS IF USING CUBES)

1 TSP THYME LEAVES

ABOUT 1½ TBSP HONEY

Preheat the oven to 220°C/425°F/Gas 7. Cut the potatoes into large chunks of roughly equal size, put them in a pan of well-salted water, bring to the boil and boil for 5 minutes.

Meanwhile, slash the top of the lamb approximately five times, about 5mm/¼in deep. Rub 1 tbsp oil all over the lamb and season well with salt and pepper. Run your fingers down the thyme sprigs, from base to tip, to scatter the leaves over the meat, reserving ½ tsp to add just before serving.

Drain the potatoes and rough up the outsides using two forks. Season with pepper and sprinkle over the flour. Toss the potatoes around in the flour.

Put a roasting pan in the oven with 2 tbsp oil for 3 minutes or so, until it is very hot. Add the potatoes and turn them in the fat, then push to the outside of the pan. Put the lamb

in the middle, return to the oven and roast for 20 minutes. Turn the oven down to 190°C/375°F/Gas 5 and continue to cook for 15 minutes per 500g/1lb 2oz for pink lamb and 20 minutes per 500g/1lb 2oz if you prefer your meat well-done. Turn the potatoes around in the fat once or twice during cooking.

Remove the potatoes and put in a serving dish, crisping them up further in the oven if necessary. Transfer the lamb to a carving board or a large serving platter. Cover loosely with foil and leave in a warm place to rest while you make the gravy.

For the gravy, put the roasting pan over a medium–low heat. Sprinkle in the flour, stir it into the fat and cook for a minute or so. Add the cider and stock and stir well to lift up all the remaining tasty bits from the bottom of the pan. Stir in the thyme leaves. Allow the gravy to bubble away until slightly thickened, then add the honey.

Warm 1 tbsp honey, then brush this over the lamb and scatter with the reserved thyme leaves – they will stick to the honey. Tip any juices that have come out of the lamb into the gravy and stir to combine.

Serve the lamb and potatoes with the gravy. The leeks with spicy pollen breadcrumbs (page 96) and/or the ginger-glazed carrots (page 93) go very well with this, along with a green vegetable.

ROAST GROUSE WITH A HONEY, BLACKBERRY AND WHISKY SAUCE

Grouse is one of the best game birds because its dark meat has the savour of the moorland where the birds live. The season starts in late summer and coincides both with blackberries and the ling heather that purples the moors and produces one of my favourite honeys. You shouldn't mess too much with grouse and it works best simply roasted, but in this recipe the blackberries add a beautiful pink colour to the gravy while the honey and whisky bring a sweet smokiness.

This honeyed gravy also works well with roast partridge or even a roast chicken that is served as summer turns to autumn, when the leaves are still on the trees but you want a proper Sunday lunch or roast supper as the days turn colder.

SERVES 2

2 GROUSE

A LITTLE BUTTER

SEA SALT AND FRESHLY GROUND BLACK PEPPER

2 THIN RASHERS OF SMOKED STREAKY BACON

1 TSP HONEY

BLACKBERRY AND WHISKY SAUCE

1 TSP PLAIN FLOUR

2 TBSP WHISKY OR OTHER SPIRIT

200ML/7FL OZ CHICKEN OR GAME STOCK

ABOUT 10 WILD BLACKBERRIES (OR 5–6 CULTIVATED BLACKBERRIES)

1 TBSP HONEY (HEATHER HONEY IS IDEAL)

LEMON JUICE, TO TASTE

Preheat the oven to 230°C/450°F/Gas 8. Smear a little butter over the top of each bird and season with salt and pepper. Cut each rasher of bacon in half and drape over the breasts of the birds (the fat will baste the meat and stop it drying out).

Put the birds in the oven for 10 minutes. After this time, the bacon may have crisped up and need to be taken out. If so, quickly baste the birds with fat from the pan and put back in the hot oven for another 10 minutes. Take the birds out, put on a warm plate and cover with foil and a clean tea towel to rest while you make the sauce.

Put the roasting pan over a medium–low heat and add about 1 tsp more butter if necessary. Sprinkle over the flour and stir it around so it cooks a little. Splash in the whisky and let it bubble up quickly, stirring in any tasty gunk from the bottom of the pan. Add the stock and blackberries and cook for a couple of minutes, squishing down the blackberries slightly to release their juices.

Add the honey and stir to dissolve. Let the sauce bubble away until slightly thickened. You may prefer to remove the blackberries once they have released their flavour and colour into the sauce. I tend to remove the large cultivated ones but leave the smaller wild ones in. Taste the gravy, season with salt and pepper and adjust the sweet–sour balance by adding a little more honey or a little lemon juice as necessary.

Serve the grouse traditionally with the bacon, watercress, potato crisps, some breadcrumbs fried in butter and the sauce. Or instead of the crisps serve sautéed or roast potatoes with the grouse.

FIVE-SPICE DUCK WITH A HONEY GLAZE

Honey and Chinese flavours work so well together; the sweetness and the fragrantly spicy element of the cuisine creating a particular magic. Five-spice powder is in itself a magic ingredient with five strong spices (star anise, cloves, cinnamon, Sichuan pepper and fennel) that hold each other in check without losing any of their original fragrance. Once in your cupboard, five-spice soon finds its way into dishes that are not Asian, such as lamb meatballs or roasted root vegetables, and I often reach next for a honey pot.

The method used to cook the duck ensures that it is perfectly pink. I found this trick in Bruce's Cookbook *by Bruce Poole, a distillation of many discoveries from the decades this great chef has dedicated to the kitchen.*

Serve this with potatoes of some kind — mashed, dauphinoise — or noodles or rice, and some stir-fried vegetables dressed with sesame oil, or a salad with an Asian-inspired dressing, such as vinaigrette with a little sesame oil, orange juice and honey.

SERVES 4

4 DUCK BREASTS, ABOUT 200G/7OZ EACH

2 TSP FIVE-SPICE POWDER

SEA SALT

2 TSP HONEY

Take the breasts out of the fridge 20 minutes before cooking. Slash the breasts on the diagonal four or five times, cutting just through the skin to the flesh. Rub ½ tsp five-spice power and a good pinch of sea salt into each breast. Put the duck skin-side down in a large frying pan and turn the heat on to medium–high. The fat will come out of the duck and the flesh will be on the way to being cooked rare by the time the skin is crisp and dark. The exact time will depend on your pan and the size of the breasts, but 10 minutes is about right.

Turn the duck over and cook for another 3 minutes, or 5 minutes if you want it more well done. Remove from the pan onto a warm plate. Drizzle the honey over the duck skin, spreading it with the back of a wooden spoon. Runny honey is easiest, but thicker honey will soon melt on the hot skin. Turn the breasts over on the plate so they are covered in the melting honey. Leave to rest in a warm place for 5 minutes: this is essential to get evenly pink meat.

Cut through one breast to check it is perfectly pink. If not, put the duck back in the pan for a minute or two. This shouldn't be necessary, but if so you'll need to watch that the honey doesn't burn. Serve hot.

CHICKEN WITH HONEY, LEMON AND THYME

This is one of those simple, any-day recipes that suits a family meal or supper with friends. You can vary the flavourings according to whim, replacing the thyme with perhaps 2 tsp five-spice powder, 1½ tsp garam masala or 1 tsp smoked paprika.

You need to be careful about baking honey because it can burn if left too long, but if you combine it with lemon juice and a little oil, then it can cook for a bit longer. The chicken skin should be nicely burnished by the time the chicken thighs are cooked.

SERVES 4

8 SMALL OR MEDIUM CHICKEN THIGHS, WITH SKIN AND BONE

3 TBSP HONEY (A MEDIUM OR DARK ONE SUCH AS THYME WORKS WELL)

1½ TBSP OLIVE OIL

JUICE AND FINELY GRATED ZEST OF 1½ UNWAXED LEMONS

2–3 GARLIC CLOVES, FINELY CHOPPED

A SMALL HANDFUL OF THYME STALKS

SEA SALT AND FRESHLY GROUND BLACK PEPPER

Slash each chicken thigh three times, going down to the bone. Put the chicken in a wide, shallow baking dish. Drizzle over the honey and then the olive oil. Pour over the lemon juice and scatter the zest over the chicken. Scatter over the garlic and thyme. Season with plenty of salt and pepper. Rub the mixture thoroughly into the chicken and, if you have time, cover and leave to marinate in the fridge for at least 1 hour, or overnight.

Take the chicken out of the fridge 15 minutes before cooking if it has been in there for more than a couple of hours.

Preheat the oven to 190°C/375°F/Gas 5. Line a baking sheet with baking parchment. Place the chicken thighs on top, ideally so they are not touching. Cook for around 30 minutes – after about 20 minutes, turn the baking sheet round and move the outer pieces to the centre to ensure the chicken cooks evenly.

Serve with a green salad and potatoes; you can cook potato wedges or baked potatoes in the oven along with the chicken.

CHICKEN LIVER PARFAIT WITH HONEY-HERB JELLY

A parfait is a silky smooth pâté that the French sometimes partner with a glass of Sauternes, a wine often described as honeyed. This gave me the idea for this pretty dish of a chicken liver pâté topped with a subtle, lightly set honey jelly patterned with herbs. The preparation takes a little time, but you can do it all well in advance and have it ready to put on the table.

SERVES 6

150G/5½OZ UNSALTED BUTTER, PLUS ½ TBSP
FOR THE SHALLOTS

½ TBSP OLIVE OIL

2 SHALLOTS, FINELY CHOPPED

1 GARLIC CLOVE, FINELY CHOPPED

½ STAR ANISE

400G/14OZ CHICKEN LIVERS, CLEANED, SINEWS
AND BLOODY PARTS REMOVED

1 TSP THYME LEAVES

3 TBSP MADEIRA, DRY CIDER, SHERRY OR WHITE WINE

100ML/3½FL OZ DOUBLE CREAM

1 TSP SEA SALT

A GOOD GRINDING OF BLACK PEPPER

TOAST AND ONION MARMALADE (OPTIONAL),
TO SERVE

HONEY-HERB JELLY

1 SMALL SHEET (1.6G) OF LEAF GELATINE

100ML/3½FL OZ MADEIRA, DRY CIDER OR WHITE
WINE (HALF MADEIRA AND HALF DRY CIDER OR
WINE IS EVEN BETTER)

1½ TBSP HONEY (A HERBY ONE SUCH AS THYME
WORKS ESPECIALLY WELL)

ABOUT 10 FLAT-LEAF PARSLEY LEAVES

Melt 150g/5½oz butter in a small pan and put to one side. Put the remaining ½ tbsp butter and half the olive oil in a frying pan and heat until melted. Add the shallots, garlic and star anise and cook gently, stirring occasionally, until soft, about 10–15 minutes. Remove the star anise and tip the contents of the pan into a food processor.

Add the remaining oil to the pan, turn up the heat and add the chicken livers and thyme. Cook over a high heat, turning the livers over once, until cooked through but still pink in the centre, about 3 minutes. Pour in the Madeira or other booze and allow the liquid to bubble up for 30 seconds or so.

Put the chicken liver mixture and juices into the food processor along with the melted butter, cream, salt and pepper. Whizz until smooth.

Pass the mixture through a sieve, using a spoon to press down on the mixture and another to scrape it from the bottom into a bowl. This makes a lovely, smooth texture and is worth the effort. It also means you don't need to worry about removing all the sinew from the livers. Taste and adjust the seasoning: you want it to be well seasoned because you'll eat the parfait cold and flavours quieten as they chill.

Carefully spoon the parfait into a fairly shallow dish (about 15cm/6in square, or 19cm/7½in diameter, and 7–10cm/3–4in deep) and smooth it over with the back of the spoon. Leave to chill in the fridge.

To make the jelly, soak the gelatine in cold water for 5 minutes. In a small pan, gently heat the cider, wine or Madeira with the honey. Turn off the heat. Squeeze the water out of the gelatine and stir it into the warm liquid until it dissolves. Leave to cool, ideally over ice, until the jelly has thickened and is starting to set.

Press the parsley into the top of the chicken liver parfait and carefully pour the semi-set jelly on top. You may need to press the leaves down with a spoon to keep them within the jelly, not on top. Leave to set in the fridge. Serve the parfait in its dish, for guests to spread on toast.

HONEYED CHICKEN AND AUBERGINE BIRYANI

Food stylist and writer Sunil Vijayaker worked on the photographic shoot for this book and shared some tips from his great knowledge of Indian food about how to make a biryani especially good. There are many different honeyed components in this dish but each one plays its part and overall there is less work involved than first appears — and any time and effort is certainly worthwhile. Serve with a salad or green vegetable.

SERVES 4

250G/9OZ LONG-GRAIN RICE

1 AUBERGINE

OLIVE OIL FOR SHALLOW FRYING

4 LARGE CHICKEN THIGHS ON THE BONE, WITH SKIN

1 LARGE ONION, FINELY CHOPPED

1 GARLIC CLOVE, FINELY CHOPPED

5 CARDAMOM PODS

1½ TSP CUMIN SEEDS

2 TSP CORIANDER SEEDS

500ML/18FL OZ CHICKEN STOCK

SEA SALT AND FRESHLY GROUND BLACK PEPPER

1 CINNAMON STICK, SNAPPED IN HALF

1 TBSP HONEY, TO GLAZE

3 TBSP MIXED FINELY CHOPPED MINT AND CORIANDER LEAVES

LEMON WEDGES, TO SERVE

50G/1¾OZ POMEGRANATE SEEDS

HONEY MARINADE

3 TBSP LEMON JUICE

2 TBSP HONEY

HONEYED ONIONS

1 LARGE ONION

SUNFLOWER OR VEGETABLE OIL

1 TBSP HONEY

Soak the rice in cold water for 20 minutes and then drain. Preheat the oven to 200°C/400°F/Gas 6. For the marinade, mix together the lemon and honey in a large bowl, and season to taste. Cut the aubergine into 3cm/1¼in cubes. Heat about 2 tbsp oil in a wide sauté pan over a high heat and cook the aubergine, in two or three batches, until softened

and lightly browned. Stir into the honey marinade and put to one side. Pour some more oil into the pan and brown the chicken over a medium–high heat for about 5 minutes on each side. Transfer to a baking sheet and put in the oven for 15–20 minutes, until nearly cooked.

Pour a little more oil into the pan, if necessary, and gently cook the onion and garlic for 7–10 minutes, stirring occasionally, until soft. Add the cardamom, cumin and coriander and stir around in the onion for a minute or so. Add the drained rice and stir for another minute, until the rice is well coated in the fat and spices. Pour in the stock, season with salt and pepper to taste, add the cinnamon and press the chicken into the rice. Scrunch up a piece of greaseproof paper slightly bigger than the diameter of the pan and wet it under the tap. Put it on top of the rice, cover with a lid and cook gently (ideally on a heat diffuser) for about 10–12 minutes, until the rice has absorbed the liquid. Turn off the heat and leave for 20 minutes with the lid on. Make sure the chicken is cooked through.

Meanwhile, for the honeyed onions, cut the onion in half and then into thin semi-circular slices. Heat about 3cm/1¼in oil in a deep pan over a high heat for a couple of minutes. Add the onion and cook for 3–4 minutes or so, until browned, stirring frequently. Remove from the oil with a slotted spoon and drain on kitchen paper. Put in a bowl and stir through the honey.

Transfer the rice and chicken to a big serving platter, or serve in the pan. The rice may have formed a crust on the bottom of the pan: this is a delicious part of the dish, so ease it off the base and serve a piece for each person. Warm 1 tbsp honey (10 seconds on medium–low in a small bowl in a microwave) and brush it over the chicken. Scatter the aubergines and then the herbs over the dish and squeeze over some lemon juice. Scatter the honeyed onions and pomegranate seeds over the top and serve with a lemon wedge on each plate.

MOROCCAN CHICKEN WITH TOMATO-HONEY SAUCE

Moroccan cooks have a useful trick of adding a spoonful of honey to balance out the acidity of tomatoes in a sauce, adding both sweetness and flavour. This tip is especially useful if your tomatoes aren't the reddest and ripest. I often use a little honey in pasta sauces, as well as in Moroccan dishes.

The sauce here is powered by honey combined with ras el hanout, the famous spice mix of Morocco that translates as 'top of the shop' – the pick of what the spice merchant has to offer. This fragrant mix of many spices usually includes rose petals, another flavour that works brilliantly with honey.

SERVES 4

8 CHICKEN THIGHS, WITH SKIN AND BONE

6 TOMATOES

2 GARLIC CLOVES, FINELY CUT

12 BLACK OLIVES, PITTED AND CHOPPED INTO 3–4 PIECES

1 TBSP HONEY

½ TSP SEA SALT

1 TSP RAS EL HANOUT

A SQUEEZE OF LEMON JUICE (OPTIONAL)

1 TBSP FINELY CHOPPED FLAT-LEAF PARSLEY

½ TBSP FINELY CHOPPED CORIANDER LEAVES

MOROCCAN MARINADE

1 TBSP OLIVE OIL

1½ TSP RAS EL HANOUT

1 PRESERVED LEMON, PIPS REMOVED, FLESH AND SKIN FINELY CHOPPED

1 TBSP HONEY

SMALL PINCH OF FINE SEA SALT

Mix together the marinade ingredients in a shallow dish large enough to hold the thighs in one layer. Slash each chicken thigh three times through the skin and mix with the marinade so they are well coated. If you have time, cover and leave to marinate in the fridge for 30 minutes, or up to 24 hours.

Preheat the oven to 190°C/375°F/Gas 5. Cut the tomatoes into quarters, place in a wide, shallow ovenproof dish and mix with the garlic, olives, honey, salt and ras el hanout. Tip the chicken and marinade into the dish and mix the marinade with the tomato mixture. Turn over the chicken pieces so they are on top of the sauce, skin-side up.

Bake for 30–35 minutes (or 40–45 for especially large thighs), until the chicken has browned, the flesh is cooked through to the bone and the tomatoes have collapsed and started to form a sauce. Stir the sauce to bring it together, taste and adjust the seasoning if necessary, adding a squeeze of lemon if the dish is too sweet. Scatter the herbs over the dish.

Serve with couscous or bread to mop up the aromatic sauce, or (less authentically but to good effect) baked potatoes.

HONEY-LIME SALMON WITH NOODLES

Honey thickens in the oven to make a glossy glaze. Here it is mixed with lime juice and water so that it doesn't become too thick and burnt, but lightly coats the fish as it cooks. This method works with any fish steak to make a simple healthy supper dish that is ready in 15 minutes.

Honey is a seasoning that brings out the flavour of the other ingredients and here it is balanced by sharp lime, salty soy sauce and nutty sesame oil, alongside the sesame seeds. Ready-toasted sesame seeds, available in Asian stores, are a useful storecupboard ingredient, or you can toast a batch at home by putting the seeds in a pan and stirring them over a medium heat for a few minutes.

Flat rice noodles are quick and easy, but you can use any kind of noodles you want. Buckwheat soba noodles work especially well with salmon and other kinds of fish, or use linguine or tagliatelle. This is a good dish for loading up veg on the plate. In addition to the carrots and sugar snaps, I sometimes fry sliced courgettes with garlic and shallots, or add some chopped kale to steam with the carrots.

SERVES 4
4 SALMON STEAKS, WITH OR WITHOUT SKIN

150–200G/5½–7OZ FLAT RICE NOODLES

2 CARROTS

50G/1¾OZ OR SUGAR SNAP PEAS OR MANGETOUT

1 TBSP FINELY CHOPPED CORIANDER LEAVES

½ TBSP SESAME OIL

SQUEEZE OF LIME JUICE

4 WEDGES OF LIME, TO SERVE

HONEY-LIME GLAZE
2 TBSP SESAME SEEDS

JUICE OF 1 LIME

2 TBSP WATER

2 TBSP HONEY (A LIGHT ONE SUCH AS ORANGE BLOSSOM OR WILDFLOWER IS BEST)

1 TBSP JAPANESE OR LIGHT SOY SAUCE

PINCH OF CHILLI FLAKES (OPTIONAL)

Preheat the oven to 190°C/375°F/Gas 5. To make the glaze, first toast the seeds: put them in a frying pan over a medium heat, stirring them around for a couple of minutes, until lightly browned. Mix together all the ingredients for the glaze and leave for a few minutes so the honey dissolves, stirring occasionally.

Line a baking sheet with baking parchment. Place the salmon on top and pour over the glaze. Carefully roll the fish in the glaze to coat it well. Put it in the oven for 5 minutes, then spoon over the glaze and put back in the oven for another 5 minutes, spooning the thickening glaze over once more about a minute before the end of the cooking time.

While the salmon is cooking, cover the rice noodles with hot water and leave to soak for 10 minutes, until soft (or follow the packet instructions). Peel the carrots and cut into thin batons. Steam the carrots for 3 minutes, adding the sugar snaps after 1–2 minutes to keep some of their crunch and vitality. I like to steam the soaked noodles briefly as well to ensure they are cooked through.

Take the fish out of the oven and leave to rest for 2 minutes, spooning over the last of the glaze.

Toss the noodles with the vegetables, coriander, sesame oil and a good squeeze of lime. Divide among four bowls and place the salmon on top, spooning over any extra glaze. Serve with lime wedges.

SWEET 'N' HOT JERK CHICKEN WINGS

Honeybees aren't native to the Caribbean; they were brought over from the Old World in the early seventeenth century. Now they happily buzz around feasting on tropical plants. One of the most unusual honeys I've got is mango honey from St Lucia, sold in a little rum bottle with a home-produced bee label stuck on the outside. I've used this to make jerk chicken, but any honey will do. The nuances of a monofloral would be lost in this dish, so use a basic honey: the fundamental honey flavour comes through, as well as the shine.

Jerk is the quintessential flavour of Jamaican food. The exact ingredients vary, but it must be spicy, hot and slightly sweet. Honey provides the sweet; the fruity Scotch bonnet chilli brings the heat. The spices should certainly include allspice, from the beautiful trees that you find all over Jamaica. Plenty of nutmeg is also an essential guest at this flavour party and so is thyme. Vary your jerk according to your tastes and develop your own house special.

These wings also cook brilliantly on the barbecue. Sweet potato wedges would be a great partner.

SERVES 4–5
20 CHICKEN WINGS
LIME WEDGES, TO SERVE

JERK MARINADE
THUMB-SIZED PIECE OF FRESH GINGER, FINELY CHOPPED
2 GARLIC CLOVES, ROUGHLY CHOPPED
GREEN PARTS OF 4 SPRING ONIONS, ROUGHLY CHOPPED
¼–½ SCOTCH BONNET CHILLI (OR OTHER CHILLI), DESEEDED AND FINELY CHOPPED
1 TSP FRESH THYME LEAVES
2 TSP FINELY GRATED NUTMEG
1 TSP GROUND ALLSPICE
1 TSP SALT
3 TBSP HONEY
100ML/3½FL OZ CIDER VINEGAR
2 TBSP OLIVE OIL

Put all the ingredients for the jerk marinade in a small blender and whizz to a smooth paste. (Alternatively, chop and grate everything up as small as possible and mix together.) Put the chicken wings in a non-metallic dish and mix thoroughly with the marinade. Cover and leave in the fridge for at least 1 hour and ideally overnight.

Preheat the oven to 190°C/375°F/Gas 5 and line a baking sheet with baking parchment. Add the chicken wings, ideally with a little space between them, and spoon over the jerk marinade.

Cook the wings for 45 minutes, spooning the marinade over once or twice, until the marinade has reduced right down to a sticky coating but hasn't burnt. Serve hot or warm with a wedge of lime to squeeze over.

SMOKED FISH AND HONEYED POTATO SALAD
WITH PINK PEPPERCORNS

The strong salty flavour of smoked fish combines and contrasts beautifully with the soft sweetness of honey in this pretty salad. Other complementary elements include lime juice, perfumed pink peppercorns and parsley. The salad works equally well as a smart starter or as a lunchtime main course.

The idea came originally from a Brazilian chef, Mariana Villas-Bôas, who I came across while in pursuit of information about stingless bee honey, one of the most extraordinary honeys I've ever eaten. It has a special floral lightness and is much prized for medicinal use. Maria uses palm hearts instead of potatoes in her salad, but these can be hard to obtain and I like the way that the honey dressing is absorbed by the hot potatoes. You can also add a handful of rocket to make it even more of a dish of pretty pinks and greens.

SERVES 4 AS A STARTER OR LUNCH DISH

600G/1LB 5OZ SMALL SALAD POTATOES, PEELED

SALT

100G/3½OZ HOT-SMOKED SALMON, OR OTHER SMOKED FISH

1–2 TSP PINK PEPPERCORNS

HONEY AND LIME DRESSING

4 TBSP RUNNY HONEY (ANY LIGHT HONEY, SUCH AS ACACIA)

1 TBSP FRESH LIME JUICE (ABOUT 1 AVERAGE LIME)

3 TBSP EXTRA VIRGIN OLIVE OIL

SEA SALT

2 TBSP FINELY CHOPPED FLAT-LEAF OR CURLY PARSLEY

Cut the potatoes in half, or if necessary into pieces roughly the same size, about 6cm/2½in long. Put them in a pan of well-salted water, bring to the boil, then turn down the heat and simmer for about 8–10 minutes, until tender.

Meanwhile, make the dressing: put the honey, lime juice, oil and a small pinch of salt in a serving dish. Whisk together with a fork and leave to allow the salt to dissolve while the potatoes are cooking.

Drain the potatoes and cut them into roughly 3cm/1¼in pieces. Stir the parsley into the dressing and then toss the potatoes in the dressing so they are well coated. Flake the salmon into chunks and carefully mix into the dressed potatoes, taking care not to break up the pieces. Crush the pink peppercorns roughly in a mortar and pestle and scatter over the salad. Taste for seasoning, adding a squeeze more lime or more salt if necessary. Serve warm or at room temperature. This makes a good lunch dish alongside a green salad or some lightly steamed vegetables.

CARAMELIZED ONION TART

Onions sweeten deeply when slow-cooked, and a spoonful or two of honey enhances this quality. Grainy mustard and the strongly aromatic taste of caraway seeds act as a counterbalance. You can make this tart slightly in advance of guests arriving and serve it warm with hot potatoes and a salad. You can also bake the pastry case and cook the onions the day before and then bake the tart on the day.

SERVES 4–5

2 TBSP OLIVE OIL

5 MEDIUM OR 3 LARGE ONIONS (ABOUT 450G/1LB), FINELY SLICED

½ TSP CARAWAY SEEDS

2 TBSP HONEY

2 TBSP GRAINY MUSTARD

3 LARGE EGGS

150ML/5FL OZ FULL-FAT MILK

1½ TBSP FRESHLY GRATED PARMESAN CHEESE

PASTRY

175G/6OZ PLAIN FLOUR, PLUS EXTRA FOR DUSTING

PINCH OF SALT

100G/3½OZ COLD UNSALTED BUTTER, CUBED

1 TBSP GRAINY MUSTARD

1–2 TBSP COLD WATER

First make the pastry. Mix the flour and salt in a bowl. Rub the butter into the flour using the tips of your fingers. Add the mustard and about 1 tbsp water, or as much as you need to bring the mixture together into a ball. Wrap in clingfilm and put in the fridge for 30 minutes; this makes the pastry easier to handle and less likely to shrink in the oven. Meanwhile, heat the oil in a frying pan over a medium heat. Add the onions and cook for 10 minutes, stirring often. Turn the heat to low, stir in the caraway seeds and cook for 20 minutes, still stirring. Stir in the honey and 1 tbsp of the mustard, then turn off the heat.

Roll the pastry out on a lightly floured surface and use to line a 20cm/8in diameter loose-bottomed tart tin. Put back in the fridge for 15 minutes. Meanwhile, put a baking sheet in the oven and preheat it to 200°C/400°F/Gas 6. Line the pastry case with foil and cover with baking beans. Cook for 15 minutes, then remove the beans and cook for another 5–10 minutes to dry out the pastry centre. Take the case out of the oven and turn the heat down to 180°C/350°F/Gas 4. Spread the onions over the base. Whisk together the eggs, milk and the remaining 1 tbsp of mustard and pour over the onions. Scatter the cheese on top. Bake for about 25 minutes, or until nicely browned on top. Serve warm.

BLUE CHEESE AND FLOWER SALAD WITH HONEYED WALNUTS

The world of edible flowers was opened up to me by a visit to Maddocks Farm Organics in Devon, where Jan Billington specializes in growing organic flowers for the table. We wandered through her beds of flowers, munching away at exquisite blooms, while Jan showed me which ones the bees preferred to feast on.

Ever since, I have looked at flowers with new eyes. We eat the leaves of plants, so why not the flowers? You need to make sure you're eating ones that taste nice and aren't toxic, but once you've found a few favourites, flowers are an easy way to add fresh beauty and interest to the plate. The Scented Kitchen: Cooking with Flowers *by Frances Bissell is the best book I've found on the subject, along with* Edible Flowers *by Claire Clifton.*

The chives, borage and violas used here are all blue flowers to go with the blue cheese and are all easy to grow, whether you have a plot or a pot. Otherwise, you occasionally find such blooms in some greengrocers and the odd supermarket, but mostly you need to get in touch with someone like Jan.

SERVES 6

60G/2¼OZ WALNUT PIECES

1¼ TBSP HONEY

600G/1LB 5OZ SALAD LEAVES — MOSTLY GENTLE, SUCH AS BUTTERHEAD OR COS (ABOUT 400G/14OZ), WITH SOME STRONGER, MORE HERBAL LEAVES SUCH AS ROCKET OR MIZUNA (ABOUT 200G/7OZ)

250G/9OZ BLUE CHEESE, SUCH AS STILTON, ROQUEFORT OR BEENLEIGH BLUE, ROUGHLY CRUMBLED OR CUT INTO PIECES

BLUE FLOWERS — PERHAPS 12 BORAGE FLOWERS, 4 CHIVE HEADS (PETALS PULLED APART), 6–12 VIOLAS

HONEY DRESSING

2 TBSP HONEY (A LIGHT ONE SUCH AS WILDFLOWER, ACACIA OR ORANGE BLOSSOM IS BEST)

¼ GARLIC CLOVE, FINELY CHOPPED, PLUS THE REST OF THE CLOVE TO RUB ON BREAD

1½–2 TSP FRUIT OR CIDER VINEGAR

SEA SALT

6 TBSP MILD OLIVE OIL

To make the honeyed walnuts, preheat the oven to 180°C/350°F/Gas 4.

To make the dressing, put the honey in a small bowl, add the garlic and vinegar and season with a pinch of salt (the cheese is salty, so you don't need too much). Stir to dissolve the honey, then pour in the oil and whisk together. Taste and adjust the seasoning as necessary.

Put the walnut pieces on a baking sheet lined with baking parchment. Toast in the oven for 5 minutes, or until lightly crisp. Drizzle over the honey and turn over using 2 teaspoons to coat the nuts in the honey. Return to the oven for 2 minutes, taking care they do not toast too much and become bitter. Remove from the oven and leave to cool.

Shortly before you want to eat, wash and dry the salad leaves and put in a large bowl. Re-whisk the dressing and pour it over the leaves. Mix carefully and gently with your hands. Place on 6 plates and scatter over the cheese. Arrange the flowers and honeyed walnuts on top.

BAKED CHEESE WITH HONEY-WALNUT TOASTS

Honey and cheese is a marriage made in heaven, appreciated by food-lovers from the ancient Greeks and Romans onwards. Nuts go well with both honey and cheese and this simple version of a fondue – a whole cheese cooked in a box – brings them all together as you dip pieces of walnut toast spread with honey into the oozy melted cheese.

This is a dish for bringing out your best honey. Because the honey is spread on toast and not heated, all the glory of a monofloral or special honey will be on full display.

Most Camembert boxes are held together with staples; make sure it is one of these and not one with glue.

SERVES 4 AS A LIGHT SUPPER WITH SALAD, OR 4–6 AS A STARTER

1 WHOLE CAMEMBERT

½ GARLIC CLOVE (OPTIONAL)

A SPLASH OF DRY WHITE WINE OR KIRSCH

FRESHLY GROUND BLACK PEPPER

A SCATTERING OF FRESH THYME OR OREGANO LEAVES, CHOPPED, TO FINISH (OPTIONAL)

TO SERVE

1 SMALL LOAF OF WALNUT BREAD

ABOUT 2 TBSP HONEY

150G/5½OZ AIR-CURED HAM, SUCH AS PARMA OR SERRANO

1 SMALL POT OF GHERKINS, DRAINED

Preheat the oven to 200°C/400°F/Gas 6. Take the cheese out of its box and remove the wrapper. If you like, rub the bottom, top and sides lightly with a cut clove of garlic. Put the cheese back in the box and discard the lid. Place on a baking sheet and bake for 15 minutes, until the inside is soft and melted. You will be able to feel this by pressing the top, or cut off a small section of the top and look inside.

Meanwhile, cut thin slices of walnut bread and toast. Spread lightly with honey and put on a serving plate. Put strips of air-cured ham on another plate and put the gherkins in a bowl.

Carefully cut the top off the cheese. Pour over a little wine or Kirsch (the traditional liqueur to use in a fondue). Season with a few grinds of black pepper. If you like, you can also scatter over some chopped thyme or oregano.

Serve with the honey toasts: dip them into the cheese, or spread the oozy cheese over the toasts. Eat with the ham and gherkins. For a main course, serve with a sharply dressed green salad.

GOAT'S CHEESE, HERB AND HONEY FILO PARCELS

Put a plateful of these tasty little pastries on a table and people come to them like bees to blooms. The filo pastry is spread lightly with honey, then wrapped around a mixture of salty cheese, fresh herbs and courgette, tangy spring onion and lemon zest, and crunchy pine nuts. A sprinkling of fragrant seeds on top of the pastry completes the flavours.

Once you've got the knack of folding the parcels, they slip easily into a production line, best accompanied by music, the radio or talking to a fellow cook. I wouldn't recommend these, however, if you are in a stressed rush. Filo pastry has a way of punishing the hasty cook by ripping and generally misbehaving. Otherwise it is easy-peasy(ish).

MAKES 12 PARCELS
4 TBSP OLIVE OIL

2½ TBSP RUNNY HONEY, PLUS EXTRA FOR DRIZZLING

ABOUT 280G/10OZ FILO PASTRY (6 LARGE SHEETS)

1 EGG, BEATEN

3 TBSP SESAME, POPPY AND NIGELLA SEEDS
(USED SINGLY OR MIXED)

FILLING
50G/1¾OZ PINE NUTS, TOASTED

100G/3½OZ COURGETTE (1 MEDIUM COURGETTE),
FINELY GRATED AND EXCESS WATER SQUEEZED OUT

85G/3OZ SOFT GOAT'S CHEESE

150G/5½OZ RICOTTA CHEESE

6 TBSP MIXED CHOPPED FRESH HERBS, SUCH AS MINT,
BASIL, FLAT-LEAF PARSLEY

FINELY GRATED ZEST OF ½ UNWAXED LEMON

2 SPRING ONIONS, VERY FINELY CHOPPED, INCLUDING
GREEN PARTS

SEA SALT AND FRESHLY GROUND BLACK PEPPER

Preheat the oven to 200°C/400°F/Gas 6. Line a baking sheet with baking parchment. Whisk together the oil and honey.

For the filling, you can buy toasted pine nuts or you can toast them on a baking sheet in the hot oven for 2–3 minutes, taking great care not to burn them. (You can also toast them – less well but with less risk of burnt-nut disaster – in a frying pan over a medium heat, stirring the pine nuts for 3–4 minutes, or until patched with brown.) Leave the pine nuts to cool slightly.

Mix together the ingredients for the filling, adding the pine nuts at the end, and season well with salt and pepper to taste.

Take one filo sheet at a time and lay it on the work surface. Cut the sheet in half to get two long strips about 10–12cm/4in wide. Brush with the oil and honey mixture. Put a dessertspoonful of filling in the middle of the top of one strip. Fold the pastry over diagonally to make a triangle and as you do so flatten out the filling slightly. Fold this triangle down towards you, keeping the triangle shape. Continue folding in triangles, first one way and then the other, until you reach the end of the strip. Trim off the end to make a neat parcel and place on the lined baking sheet. Repeat with the rest of the pastry and filling.

Brush the parcels with beaten egg and scatter the seeds over the top. Bake for 10 minutes, or until golden. If you like, drizzle with more honey. Serve warm or hot, allowing two or three pastries per person, either with a salad or as part of a table of mezze.

GRILLED GOAT'S CHEESE AND CHICORY SALAD WITH HONEY AND POPPY SEED DRESSING

This recipe comes from Darren Marshall, head chef at the Island Grill in the Lancaster London hotel. The Lancaster has played a full part in the revival of urban beekeeping in London: it has hives on the roof and serves the honey in the hotel, and hosts an annual honey festival. Darren is part of the Lancaster Bee Team, which looks after the hotel's hives. This has given him the chance to create recipes that suit the very specific taste of the honey. 'Our bees produce a honey with a distinctive lime flavour from foraging in Hyde Park just across the road,' says Darren. Any fragrant light honey works well in this recipe: lime blossom or acacia honey are ideal.

The smooth, sweet-sharp dressing sets off the mixture of textures and tastes, and it looks great, with the poppy seeds and red onion flecking the white cheese, crisp chicory and darker salad leaves. Darren sometimes adds toasted walnuts as well.

SERVES 4 AS A MAIN COURSE

1 TSP VEGETABLE OR SUNFLOWER OIL

2 RED ONIONS, FINELY SLICED

4 HEADS OF CHICORY, IDEALLY 2 WHITE AND 2 RED

1 HANDFUL OF BABY SPINACH

1 HANDFUL OF ROCKET

400G/14OZ GOAT'S CHEESE IN A SMALL, WHITE-RINDED LOG (SUCH AS 2 X RAGSTONE LOGS)

TOASTED FLATBREAD, TO SERVE

HONEY AND POPPY SEED DRESSING

2 TBSP HONEY (SEE INTRODUCTION)

1 TSP POPPY SEEDS

4 TBSP EXTRA VIRGIN OLIVE OIL

3–4 TBSP WHITE WINE VINEGAR

½ TSP DIJON MUSTARD

SEA SALT AND FRESHLY GROUND BLACK PEPPER

Heat the oil in a pan over a medium–low heat and gently cook the onions for 15 minutes, until soft and sweet, stirring occasionally at first and then more frequently towards the end. Leave to cool.

Trim the bottoms off the chicory, then remove the leaves one by one. You will need to cut the stalk again every two or three layers. Slice the leaves lengthways. Darren likes to cut them very thin, but you can cut them thicker for a different texture. Put in a wide salad bowl.

Slice the spinach and add to the chicory. Add the rocket leaves, picking off any large stalks.

Preheat the grill to high.

To make the dressing, put all the ingredients in a small bowl and whisk together. Taste and adjust the seasoning. This dressing is on the sharp side; so, use more or less vinegar, according to taste.

Using a warm knife, cut off the rind from both ends of the goat's cheese. Cut the cheese into slices about 2cm/¾in thick, place on a baking sheet and put under the grill for around 30 seconds. Make sure you do not over-melt the cheese; you are just warming it through. Leave the slices to cool slightly.

Re-whisk the dressing and pour over the salad, mixing thoroughly to coat the leaves. Place the grilled cheese on top, scatter over the cooked onions and serve with toasted flatbread.

BEETROOT SALAD WITH HONEYED APPLES

I love the look of this dark-purple salad, flecked with sweet-and-sour red onion in a honey and cider vinegar dressing and mixed with honeyed apples. It is a lovely dish on its own or goes brilliantly with a pâté and a green salad. You can add some good canned tuna, anchovies, walnuts or soft goat's cheese to turn it into a main course. Cooked beetroot can also be dressed simply in a mixture of honey and yogurt to make a bright pink dish.

SERVES 6–8 AS A SIDE SALAD

4 BEETROOT

2½ TBSP OLIVE OIL

3 TBSP HONEY (A LIGHT ONE SUCH AS ACACIA OR ORANGE BLOSSOM IS BEST)

2–3 TBSP CIDER VINEGAR

SEA SALT AND FRESHLY GROUND BLACK PEPPER

1 RED ONION, FINELY SLICED

3 TBSP FINELY CHOPPED DILL

2 APPLES, IDEALLY A SWEET-SHARP VARIETY SUCH AS COX

You can buy ready-cooked beetroot — good greengrocers often make a batch — but make sure they haven't been stored in vinegar. If your beetroot are raw, start by cooking them. Preheat the oven to 200°C/400°F/Gas 6. Wrap each beetroot individually in foil and bake for 1¼–1½ hours, until tender all the way through when you slide in a sharp knife. They take a bit longer to be truly tender than you'd think.

Meanwhile for the dressing, put 2 tbsp of the oil into a serving bowl and use the oily spoon to measure out 2 tbsp honey so it slips easily off the spoon. Add the vinegar, starting with 2 tbsp (you can add more later if necessary). Whisk the dressing together and season with salt and pepper. Mix the onion slices with the dressing. Cover and leave to marinate for at least 1 hour. When the beetroot are cooked, remove the foil and leave until cool enough to handle, then peel off the skins and trim off the stalks. Cut into quarters and then cut each quarter into thin slices. Add to the onion, along with the dill. Stir well.

Cut the apples into quarters and remove the cores. Cut the quarters into slices, around 1cm/½in thick (five to seven slices per quarter apple). Put the remaining ½ tbsp oil in a frying pan over a medium heat. Add the apple slices and cook for a minute. Season with salt and pepper and drizzle over the remaining 1 tbsp honey. Turn the apple slices over and let the honey bubble up for 30 seconds to a minute. Scatter the honeyed apple slices over the beetroot mixture and stir in. Taste the salad and sprinkle over a little more vinegar if you prefer a more sharp-sweet version; it also depends on whether you are serving the salad with another dish. When serving with a rich pâté, I like to make it slightly sharper. Season with a grind of pepper and serve warm or at room temperature.

HONEY-CARAMELIZED FENNEL SOUP

Fennel and honey are good companions, the herbal, aniseedy twang of the fennel balanced by the sweetness of the honey in this fragrant sweet-and-sharp soup. Fennel pollen is a cultish ingredient that crops up in trendy cocktails, sweets and savoury dishes and is particularly popular with Italian chefs. In Italy the pollen has long been gathered by hand from the flowers, rather than coming from bee hives. I home-harvested some from the statuesque fennel flowers in the garden, an idea that occurred after I started to look at plants from a bee's eye view. If they can gather, why can't I? Fennel pollen is less sweet than other commercial pollens and makes an interesting garnish for the soup. Standard pollen is widely available in health food shops and also works well. If you can't find either, don't fret; this soup is beautiful and delicious just with a swirl of cream.

SERVES 4

1 MEDIUM–LARGE FENNEL BULB, TRIMMED (ABOUT 140G/5OZ TRIMMED WEIGHT), ROUGHLY CHOPPED AND ANY FRONDS RESERVED

2 TBSP OLIVE OR SUNFLOWER OIL

SEA SALT AND FRESHLY GROUND BLACK PEPPER

1–1½ TBSP HONEY (A DARK HONEY SUCH AS THYME, HEATHER OR CHESTNUT WORKS ESPECIALLY WELL)

1 TSP SHERRY VINEGAR OR WHITE WINE VINEGAR

1 ONION, FINELY CHOPPED

1 CELERY STICK, FINELY CHOPPED

1 GARLIC CLOVE, FINELY CHOPPED

1 POTATO (ABOUT 140G/5OZ), PEELED AND FINELY CHOPPED

900ML/32FL OZ CHICKEN STOCK (A CUBE IS FINE BUT BEWARE OF OVERSALTINESS)

JUICE OF ¼–½ LEMON

TO GARNISH

SINGLE CREAM

FENNEL FRONDS AND POLLEN (OPTIONAL)

Preheat the oven to 200°C/400°F/Gas 6. Put a piece of baking parchment on a baking sheet and fold it up at the edges so that it will contain the melted honey. Lay the fennel on top and toss with 1 tbsp of the oil and a good seasoning of salt and pepper. Use the oily spoon to measure out the honey and spoon it over the fennel. Put the baking sheet in the oven for a couple of minutes until the honey melts, then remove from the oven and stir the honey into the fennel.

Cook for 10–15 minutes, or until the fennel is starting to caramelize lightly at the edges and the honey has darkened, checking after 7 minutes and stirring the fennel around in the honey. Sprinkle the vinegar over the fennel and stir it around again.

While the fennel is cooking, heat the remaining oil in a saucepan over a low heat, add the onion, celery and garlic and cook for about 10 minutes, stirring occasionally, until softened. Add the potato and cook for 2–3 minutes to soften slightly.

Add the fennel and its honeyed juices to the pan. Pour in the stock, bring to the boil, then cover and simmer for 15 minutes, or until the vegetables are soft. Leave to cool slightly, then whizz in a blender.

Taste and stir in enough lemon juice to balance the sweetness of the soup, remembering that the cream will mellow out the flavours. Ladle the soup into bowls and decorate with a swirl of cream. If you have them, scatter over a few fennel fronds. A scattering of pollen also looks wonderful on this soup.

SNACKS, SIDES AND SAUCES

CHORIZO WITH WINE AND HONEY

The glory of honey is the way it makes even very simple food taste and look special. Natalie Thomson, the home economist on the photographic shoot for this book, shared her inspired idea of crisping up slices of chorizo in a frying pan and giving them a sticky coating of red wine and glossy honey. I've poured in a bit more wine than Natalie uses, in order to make more of a dark shiny sauce, but both ways are delicious.

You don't need runny honey – any kind quickly melts in the hot pan – and I wouldn't use my best honey here: the chorizo overpowers any subtleties of flavour.

SERVES 4–6

150G/5½OZ SPICY CHORIZO IN ONE PIECE (READY-TO-EAT, NOT COOKING CHORIZO)

150ML/5FL OZ RED WINE

2 TSP HONEY

Cut the chorizo into 1cm/½in slices on the diagonal. Put the slices in a large frying pan and cook over a medium–low heat for about 3 minutes, or until the chorizo starts to release its fat.

Turn the chorizo slices over and cook for another 2 minutes. Turn over once more to brown for 30 seconds.

Pour the red wine into the pan and quickly drizzle in the honey. Stir everything together for about 30–60 seconds, until the sauce is reduced and coats the chorizo in beautiful, mahogany-coloured goo. Serve hot or warm.

STICKY SESAME SAUSAGES

The idea for these honey-glazed sausages comes from Livvy Cawthorn, an award-winning cook who runs a bed and breakfast (www.chainbridgehouse.co.uk) next to the historic Union Chain Bridge, which links England and Scotland across the River Tweed in Northumberland. Her neighbours are the Robson family at the Chain Bridge Honey Farm (www.chainbridgehoney.co.uk), and Livvy serves their honey at her breakfast table and also uses it in cooking.

The Robsons' bees produce wildflower honey and heather honey, including heather honeycomb. The honey farm is run by third-generation beekeeper Willie Robson, his wife Daphne, son Stephen (also a beekeeper) and daughters Heather and Frances. Chain Bridge Honey Farm is a great place to visit, with a café on a vintage bus and a visitor centre and shop swarming with bee- and honey-related exhibits and treats, full of life and the warm scent of honey.

SERVES 4 — OR SCALE UP FOR PARTIES
2—3 TBSP SESAME SEEDS
8 MEATY SAUSAGES OR 12 CHIPOLATAS
2 TSP HONEY

To toast the seeds, put them in a frying pan over a medium–high heat, stirring them around for a couple of minutes, until lightly browned.

Preheat the oven to 190°C/375°F/Gas 5. Line a baking sheet with baking parchment and lay the sausages on top, leaving a little space between them so they cook evenly. Cook in the oven – 25–30 minutes for chipolatas and about 40–45 minutes for larger sausages – until starting to brown. Spoon over the honey and quickly turn the sausages around in the goo so they are coated all over. Scatter the seeds over and turn the sausages around again.

Put the sausages back in the oven and cook for another 10 minutes. Remove from the oven and leave to cool for 5 minutes or so before serving hot; they're also great warm or cold.

COURGETTES WITH PINE NUTS, HONEY AND BASIL

Courgettes, pine nuts and honey are a trio made for one another. The soft oiliness of the courgette, the crunchy toasted nuts and the sweet glistening honey work a certain magic. You can griddle the courgettes and give them an attractive stripy appearance and chargrilled flavour, but I find the softer, richer texture of fried courgettes, halfway between a vegetable and a sauce, best accompanies simply grilled or fried meat or fish.

SERVES 4

3 MEDIUM—LARGE COURGETTES

50G/1¾OZ PINE NUTS

1½—2 TBSP OLIVE OIL

2 GARLIC CLOVES, FINELY CHOPPED

1 TSP FINELY GRATED LEMON ZEST (USE UNWAXED FRUIT)

2 TBSP LEMON JUICE

12 BASIL LEAVES

¼—½ TSP SEA SALT FLAKES

FRESHLY GROUND BLACK PEPPER

1½ TBSP HONEY

Trim off the stalk ends of the courgettes and cut each one in half. Stand each half on its end and slice downwards to get long slices about 1cm/½in thick. Put to one side.

Toast the pine nuts in a dry frying pan over a medium heat for 3—4 minutes, or until patched with brown, stirring occasionally to start with and then more often after 2 minutes. Take care they do not burn. Tip the pine nuts into a wide shallow serving dish. Return the pan to the heat and pour in 1 tbsp oil. Cook a third of the courgettes over a medium—high heat until they are patched with brown on both sides and floppy in texture. Transfer to the serving dish and, if necessary, add another ½ tbsp oil to the pan. Cook another third of the courgettes in the same way. Add ½—1 tbsp more oil to the pan and the rest of the courgettes. When you turn these over to brown the second side, add the garlic and lemon zest.

Turn off the heat and tip everything into the serving dish. Pour the lemon juice into the hot pan, stir quickly, and then pour into the serving dish.

Roughly tear up the basil leaves and scatter over the courgettes. Sprinkle with sea salt and grind over plenty of pepper. Carefully turn the courgettes over, using two spoons, so they are combined with the seasonings, basil and pine nuts. Drizzle the honey over the top. Serve hot, warm or at room temperature.

GINGER-GLAZED CARROTS

Carrots are naturally sweet and so this dish counterbalances the extra sweetness of the honey with the savoury tang of rosemary, a little ginger and a dash of soy sauce.

Slow-cooking vegetables in very little liquid is both delicious and practical. They are ready as soon as they are tender but can be left for an hour or more over a low heat, getting sweeter and more unctuous as time goes on (add a little water from time to time to prevent burning). It's a forgiving, take-it-easy technique that's useful when you have guests and time melts somewhat, especially when cooking Sunday lunch.

The soy sauce and ginger suggest these carrots as an accompaniment to dishes from both East and West. Good with five-spice duck (page 64), they are also superb with any roast meat and turn a plain piece of ham or grilled fish into a fine supper.

SERVES 6

500G/1LB 2OZ CARROTS, PEELED

20G/¾OZ UNSALTED BUTTER

5CM/2IN PIECE OF FRESH GINGER, PEELED AND FINELY GRATED

1½ TBSP HONEY

2 SPRIGS OF ROSEMARY

SEA SALT AND FRESHLY GROUND BLACK PEPPER

½ TBSP SOY SAUCE

Cut the carrots into big chunks, 6–8cm/2½–3in long, on the diagonal. Put in a heavy-bottomed saucepan and add the butter, ginger, honey, rosemary, a big splash of water, salt and pepper.

Scrunch up a piece of greaseproof paper slightly bigger than the diameter of the pan and wet it under the tap. Tuck the paper around the carrots and cover with a lid. Cook over a very low heat for at least 30 minutes and ideally longer – 1 hour or more – but check occasionally that the bottom isn't burning. If necessary, add a tablespoon or so of water. Or you can turn off the heat and leave the carrots in the pan with the lid on until you are ready to eat.

Just before serving, stir a little soy sauce into the buttery honeyed carrots to offset their sweetness with a little savoury edge.

ROMAN HONEY MUSHROOMS

The cooking of ancient Rome was full of honey. Honey sweetened food before sugar came into use and also counterbalanced the strong, salty flavour of the garum, or fish sauce, that was much used at this time. I discovered this recipe in The Classical Cookbook *by Andrew Dalby and Sally Grainger, a great journey through the food of the past, with recipes updated for the modern kitchen. I love strong flavours and am happy to follow honey into another kind of cooking that feels so different. The past is another country and this is a taste of somewhere worth exploring.*

The sweet-salty combination is made even punchier by the herb lovage (celery leaf is an alternative if you can't find lovage) and plenty of black pepper. To measure out the black pepper, loosen the top of a grinder and grind the pepper onto a piece of paper, then fold this in half to slide the pepper into a spoon measure.

Such is the spicy pungency of the dish that it is a relish as much as a side dish, good to jazz up cold meats or to eat with a piece of fish and rice, or in a baked potato.

SERVES 4–6

250G/9OZ CHESTNUT MUSHROOMS

1 TBSP OLIVE OIL

½ TSP FISH SAUCE

1 TBSP HONEY (A MEDIUM OR DARK ONE SUCH AS THYME OR GREEK HONEY IS BEST)

1 TBSP FINELY CHOPPED LOVAGE OR CELERY LEAF

½ TSP FRESHLY GROUND BLACK PEPPER

Wipe the mushrooms and cut off any manky parts of the stalks and caps. Slice finely.

Heat the oil in a large frying pan over a medium–high heat. Add the mushrooms and cook for about 5–6 minutes, stirring occasionally, until they have released most of their liquid.

Stir in the fish sauce, honey, lovage and pepper. Cook for 1–2 minutes, stirring from time to time, until the honey has reduced to stickiness and glazes the mushrooms. Serve hot, warm or at room temperature.

SPICED RED CABBAGE

Spices and honey complement each other and bring warmth and flavour to a meal. Red cabbage here gets an extra dimension with sweet honey combined with caraway, juniper, cloves, cinnamon and black pepper. The exact spicing is down to personal preference. I'm often intrigued by the way two spices alter each other, then another one makes them all change again; and so on, ad infinitum. *No wonder some cooks become obsessed by spice blends: I am one of them.*

This fragrant side dish is even better when made a day or two in advance. It is a useful accompaniment to a winter supper of baked ham or sausages.

SERVES 8

1 RED CABBAGE, ABOUT 1KG/2LB 4OZ

30G/1OZ UNSALTED BUTTER

2 COOKING APPLES

½ TBSP CARAWAY SEEDS

1 TBSP JUNIPER BERRIES, LIGHTLY CRUSHED

3 CLOVES

1 CINNAMON STICK

3 TBSP CIDER VINEGAR

200ML/7FL OZ WATER

2 TBSP CRANBERRY OR REDCURRANT JELLY

3 TBSP HONEY

1 TSP SEA SALT

FRESHLY GROUND BLACK PEPPER

Cut the cabbage into quarters. Cut out the hard white core of each quarter, then finely slice the cabbage. Melt the butter in a large pan over a medium heat and add the cabbage. Stir well to mix with the butter, then cover with a lid. Let the cabbage wilt, stirring occasionally, for about 5 minutes while you prepare the apples.

Quarter and core the apples (no need to peel) and cut into 1cm/½in pieces. Add to the pot along with the rest of the ingredients, ending with a good grinding of pepper. Bring to the boil, stirring occasionally, then turn the heat down to low, cover with the lid and simmer for 2 hours, or until the cabbage is soft and the flavours have combined to a honey-spiced sweet-and-sour.

Turn up the heat to drive off the remaining liquid, stirring often to ensure the cabbage doesn't catch on the bottom of the pan. Remove the cinnamon stick and also the cloves and juniper berries if you spot them. Serve hot, or store, covered, in the fridge for up to 3 days and reheat before serving.

LEEKS WITH SPICY POLLEN BREADCRUMBS

Leeks absorb the sweetness of honey and the sweet tartness of cider to make a side dish that is good enough to take centre stage. The leeks cook slowly in their own juices, the flavours enhanced by just a small amount of honey. Then the dish gets another layer of bee-magic with a scattering of crisp spicy breadcrumbs enriched with pollen. You don't need to make the breadcrumbs — the leeks are good without them — but they look and taste great.

With some soft goat's cheese or feta scattered over the top, these leeks make a good vegetarian supper. Otherwise, I love them alongside lamb, pork or fish.

SERVES 4

4 LEEKS, TRIMMED AND CLEANED

20G/¾OZ UNSALTED BUTTER

2 BIG GARLIC CLOVES, FINELY CHOPPED

1 TSP HONEY (MEDIUM OR DARK IS BEST)

3–4 TBSP DRY CIDER

SEA SALT AND FRESHLY GROUND BLACK PEPPER

FOR THE SPICY BREADCRUMBS (OPTIONAL)

20G/¾OZ UNSALTED BUTTER

½ TBSP OLIVE OIL

4 TBSP FRESH BREADCRUMBS

1 SMALL GARLIC CLOVE, FINELY CHOPPED

PINCH OF CHILLI FLAKES

¼ TSP FINELY GRATED ORANGE ZEST

1 TBSP FINELY CHOPPED FRESH PARSLEY

2 TSP POLLEN

Lay the leeks on a chopping board and cut in half lengthways. Then cut across the middle of each half, so each leek produces 4 long, chunky semi-circular pieces.

Put the leeks, cut-side down, into a large, heavy-bottomed pan. Dot with the butter and scatter over the chopped garlic. Drizzle over the honey, pour in the cider and season with salt and pepper.

Scrunch up a piece of greaseproof paper slightly bigger than the diameter of the pan and wet it under the tap. Put the paper over the leeks, pushing it down the edges so they are well covered. Put a lid on the pan. Bring just to the boil and immediately turn down the heat and leave to cook very slowly. The leeks will be tender after 30 minutes but are best left for 1 hour or more. Check occasionally that the bottom isn't burning and add a little more cider or water if necessary — it shouldn't be if the heat is very gentle.

While the leeks are cooking, prepare the breadcrumbs. Melt the butter with the oil in a frying pan over a low heat. Add the breadcrumbs, garlic, chilli and orange zest and season with salt and pepper. Fry gently for 2 minutes, until beginning to crisp and brown, stirring often, adding the parsley after 1 minute. Turn off the heat and put to one side.

When the leeks are cooked, put them in a serving dish and scatter over the breadcrumbs. Sprinkle the pollen over just before serving.

HONEY-ROASTED ROOTS

Just a little honey adds glistening glamour to plain roots. It's good to go for a mixture of colours and textures and a slight variation in shape among your roots, while keeping the pieces roughly the same size so they cook evenly. Besides the vegetables suggested here, parsnips and turnips are also great with honey.

This simple side dish is easy to scale up. Allow around 300g/10½oz of roots per person and take care not to overcrowd the baking sheet so that the vegetables roast evenly in a single layer.

SERVES 4

2 MEDIUM BEETROOT, ABOUT 300G/10½OZ UNPEELED

3 MEDIUM–LARGE POTATOES, ABOUT 375G/13OZ UNPEELED

1 LARGE CARROT, ABOUT 175G/6OZ UNPEELED

½ BUTTERNUT SQUASH, ABOUT 375G/13OZ UNPEELED

2 TBSP OLIVE OR SUNFLOWER OIL

¾ TSP SEA SALT FLAKES

FRESHLY GROUND BLACK PEPPER

3–4 SPRIGS OF THYME

1 TBSP HONEY, PLUS 1 TSP TO FINISH

1 TSP BALSAMIC VINEGAR (OPTIONAL)

Preheat the oven to 190°C/375°F/Gas 5. While the oven is heating up, prepare the vegetables. Peel the beetroot, cut in half and then cut each half into 3 wedges. Peel the potatoes (or give them a good scrub) and cut into 6 pieces each, of a similar or slightly larger size than the beetroot. Peel the carrot and cut into 8 chunks, at an angle. Peel the squash (or leave unpeeled if you like the chewy skin) and cut crossways into slices about 2cm/¾in thick and then cut these into pieces about 3–4cm/about 1¼in long.

Put the vegetables in a single layer on a large baking sheet. Drizzle the oil over them, sprinkle with the salt, grind over plenty of pepper and run your fingers down the thyme sprigs to scatter the leaves over the vegetables. Spoon over the honey, then use your hands to mix the ingredients together thoroughly.

Roast the vegetables for about 45 minutes to 1 hour, turning them over with a large spoon after about 30 minutes.

Spoon over 1 tsp honey and the balsamic vinegar, if you like. Taste and add more salt and pepper if necessary. Serve hot.

HONEY MAYONNAISE

As well as texture and appearance, a sauce is most of all about the way its flavours partner whatever else is on the plate. The honey in this luscious mayonnaise balances out salty foods such as ham or adds a rich note to the freshness of a crisp lettuce leaf. I especially love this mayo with cold chicken in a salad or sandwich.

Homemade mayonnaise is one of those recipes that looks like scary hard work but is really no big deal once you've done it a few times. Whisk up a small bowlful once a fortnight for a couple of months one summer and it will ever afterwards be a simple 5-minute task. It's a skill worth acquiring: homemade mayonnaise is really a different sauce to the bought kind, useful as these are, and makes a lovely meal out of simple ingredients.

Don't use too strong an oil, such as extra virgin olive oil, or it will dominate the mayonnaise in a slightly raspy way. And make sure your eggs are at room temperature before you start because this helps to form the emulsion.

SERVES 6–8

200ML/7FL OZ SUNFLOWER OR VEGETABLE OIL

100ML/3½FL OZ MILD OLIVE OIL

2 EGG YOLKS, AT ROOM TEMPERATURE

1 TSP DIJON MUSTARD

LARGE SQUEEZE OF LEMON JUICE

SEA SALT AND PEPPER (WHITE OR BLACK)

1 TBSP HONEY (LIGHT OR MEDIUM, SUCH AS ACACIA OR WILDFLOWER)

ABOUT ½ TSP WHITE WINE VINEGAR

1 TSP FINELY CHOPPED TARRAGON OR 1 TBSP CHOPPED BASIL (OPTIONAL)

Pour the oils into a measuring jug. Put the egg yolks in a steep-sided mixing bowl with the mustard, lemon juice and a pinch of sea salt. Beat well with a hand-held electric whisk, then start adding the oil very gradually, a short dribble at a time, beating each addition into the yolks before dribbling in some more.

After you've added about half the oil, you can start adding it faster because the mayonnaise is now more stable. Don't pour it all in at once, however; make sure the oil is incorporated into the emulsion as you go. Mayonnaise splits if the yolk doesn't emulsify with the oil. If this happens, whisk in another egg yolk and slowly add another 150ml/5fl oz oil (two-thirds sunflower and one-third mild olive oil).

Whisk in the honey, taste, and add the vinegar or more lemon juice. Adjust the seasoning, adding more salt and also pepper (white, say purists; I tend to have black to hand). Stir in the chopped tarragon or basil, if using.

HONEY SAUCE VIERGE

A spoonful or so of honey adds depth and texture to this simple French modern classic. A general 'all-purpose' sauce vierge — usually served with fish or shellfish — might include basil, chervil and chives. I also like tarragon, though not too much, and flat-leaf parsley suits meat such as pork or a steak. Marjoram is another good choice, especially with pork, and particularly apposite to honey because it is one of the bees' favourite plants.

A pork chop, a piece of chicken or white fish, chargrilled sliced aubergine or a soft goat's cheese salad are the perfect canvas to best display the colours and flavours of this fragrant, summery sauce. It also makes a great alternative 'gravy' for summer roast chicken, especially when it is served cold.

The one piece of faff involved in an otherwise quick recipe is skinning and deseeding the tomatoes. I don't normally agree with this, since so much of the flavour of a tomato lies in these parts. But it feels right to use chic little cubes of tomato flesh for a sauce that is all elegance. You can spread the scooped-out seeds and their tasty jelly on bread and revel in memories of squishy tomato sandwiches on hot summer days.

SERVES 6–8

2 VERY RIPE TOMATOES

100ML/3½FL OZ EXTRA VIRGIN OLIVE OIL

JUICE OF 1 LEMON

1½ TBSP HONEY (A LIGHT ONE SUCH AS WILDFLOWER OR ACACIA IS BEST)

1 SMALL GARLIC CLOVE, VERY FINELY CHOPPED OR GRATED

6 CORIANDER SEEDS, CRUSHED

3 TBSP MIXED SOFT HERBS, SUCH AS CHIVES, BASIL AND CHERVIL IN EQUAL PARTS

SEA SALT AND FRESHLY GROUND BLACK PEPPER

First, skin and deseed the tomatoes. Put them in a small bowl, cover with just-boiled water and leave for 15–30 seconds. Cut a slit in the bottom of each tomato and peel off the skin. If it doesn't come straight off, put the tomato back in the hot water for another 10–15 seconds. Cut out and discard the top of the core of each tomato and then cut into quarters. Scoop out the seeds and centre so you are left with crescents of flesh. Cut these into dice.

Put the oil in a small pan and whisk in the lemon juice and honey. When emulsified, add the remaining ingredients and season with salt and pepper to taste. Leave to infuse for at least 30 minutes.

Use on the same day or store in the fridge, covered, for a couple of days, but bring to room temperature before serving.

HONEY BARBECUE SAUCE

The challenge with a honey barbecue sauce is to make it tangy without losing the lovely warm depth of the honey. This recipe plays with the classic barbecue combination of spicy, sweet, sharp and salty. There's chilli for spicy heat, honey and pineapple juice for sweetness, lime for sharpness and Worcestershire sauce or fish sauce for saltiness. All the ingredients are balanced so that none dominates the honey.

This is an intense sauce best used as a glaze or in small quantities as a sauce for barbecued burgers, chicken, chops, sausages, hot dogs or griddled vegetables. It also adds an extra dimension to oven-cooked chicken or sausages for a quick supper. Cook the meat until nicely browned, then brush on some of the sauce for the last 5–10 minutes of cooking time.

SERVES 6–10

3 TBSP HONEY (A STRONG, DARK ONE SUCH AS THYME IS GOOD)

1½ TBSP DIJON MUSTARD

1½ TBSP TOMATO KETCHUP

2 TBSP LIME JUICE (ABOUT 2 LIMES)

125ML/4FL OZ PINEAPPLE JUICE

1 TSP WORCESTERSHIRE SAUCE OR FISH SAUCE

2 GARLIC CLOVES, FINELY CHOPPED

A SMALL AMOUNT OF FINELY CHOPPED CHILLI (DON'T OVERDO IT)

1 TSP FINELY GRATED FRESH GINGER

1½ TBSP RUM

1 TBSP VEGETABLE OIL

SEA SALT AND FRESHLY GROUND BLACK PEPPER

Put all the ingredients in a small pan over a medium heat and stir until they start to amalgamate. Turn up the heat and bring to the boil. Boil for 6–8 minutes, taking the pan off the heat after 5 minutes and letting the bubbles subside to check the consistency of the sauce. Check every minute or so, until the sauce reduces and thickens to a consistency slightly looser than tomato ketchup. It will thicken considerably as it cools.

Leave to cool and either use it that day or store in the fridge in a sealed jar or covered small bowl for up to 2 weeks.

SALTED HONEYSCOTCH SAUCE

Honey is too rich to eat on its own by the spoonful, but a salty butterscotch sauce is another matter. It's hard to believe that this is so easy to make — and to eat. Outstanding on vanilla ice cream, this delectable sauce is also wonderful with fruit and cream: I especially like it with stewed plums or pears, clotted cream and a scattering of nuts in the autumn, and single cream and fresh raspberries in the summer. The name of this sauce was coined by my friend and fellow food writer Jenny Linford. It is now one of her family favourites, usually served with roast or poached pears and vanilla ice-cream and eaten warm or when it is slightly thicker at room temperature.

Honey on its own has a slight mineral-salty tang, so go easy on the salt and use flaky sea salt rather than table salt for a gentler and greater flavour.

SERVES 6–8

50G/1¾OZ UNSALTED BUTTER

50G/1¾OZ SOFT BROWN SUGAR

50G/1¾OZ HONEY

125ML/4FL OZ DOUBLE CREAM

ABOUT ¼ TSP SEA SALT FLAKES

Put the butter, sugar, honey and cream in a saucepan over a gentle heat and stir until they have melted together.

Bring to the boil and bubble — not too hard — for a minute or so to thicken the mixture slightly. Add salt to taste. Serve warm or hot.

HONEY POPCORN

Honey is a superb alternative to sugar as a coating for popcorn and turns a snack into something special. You can also play around with salty-sweet flavours, using honey and spices such as chilli blends and five-spice powder.

Flavoured popcorn is expensive to buy — the price in cinemas is eye-popping! — so homemade is the way to go; it costs pennies and takes just a few minutes.

SERVES 2 IN FRONT OF A FILM

30G/1OZ SALTED BUTTER (OR UNSALTED PLUS A SMALL PINCH OF SEA SALT)

2 TBSP HONEY (A LIGHT ONE SUCH AS ACACIA OR ORANGE BLOSSOM IS BEST)

1 TBSP SOFT BROWN SUGAR

1 TSP GROUND CINNAMON

½ TBSP VEGETABLE OIL

50G/1¾OZ POPPING CORN

Melt the butter in a small saucepan with the honey, sugar and cinnamon over a low–medium heat, stirring often, until the sugar dissolves. Bubble for about 30 seconds to thicken slightly. Put to one side.

Pour the oil into a large saucepan and place over a medium heat for about 30 seconds. Add the corn and cover with a lid. Leave until you start to hear popping sounds. Shake the pan, with the lid on, and shake again every 30 seconds, until the popping stops. Leave for another 30 seconds.

Pour the buttery mixture over the popcorn and use two spoons to combine thoroughly. Tip the flavoured popcorn into a serving bowl. This is best eaten immediately or within a few hours, but will keep for a day or so in an airtight container.

PARMESAN, THYME AND HONEY BISCUITS

A very superior form of cheesy biscuit, this is an adaptation of a family favourite. My mother first made them to serve warm at the end of a dinner party, and they've since become a hallmark of an Ellis special occasion, be it birthday, Christmas drinks, book launch or wedding.

As well as being a foil to the salty cheese, honey is used for its colour. Combined with the egg yolk glaze, it gives a double gold effect.

Thyme is a herb that has strong associations with honey; wild thyme is a source of one of the most famous and delicious honeys in the world, Greek mountain honey. The taste of this honey summons up thoughts of warm nights on a Greek island, eating, talking and drinking cold wine.

MAKES ABOUT 20

100G/3½OZ PLAIN FLOUR, PLUS EXTRA FOR DUSTING

1 TSP FRESH THYME LEAVES (OR ½ TSP DRIED)

75G/2¾OZ COLD UNSALTED BUTTER, DICED

50G/1¾OZ PARMESAN CHEESE, COARSELY GRATED

50G/1¾OZ STRONG CHEDDAR CHEESE, COARSELY GRATED

1 LARGE EGG, SEPARATED

PINCH OF FINE SEA SALT

ABOUT 1 TBSP HONEY (MEDIUM OR DARK)

Put the flour in a mixing bowl and stir in the thyme. Rub the butter into the flour, using your fingertips, until the mixture resembles breadcrumbs. Stir the grated cheese into the crumbs. Add the egg white and use a table knife to roughly stir it into the mixture, and then use your hands to bring the crumbs together into a ball. Wrap the ball of dough in clingfilm and chill for 15 minutes.

Meanwhile, preheat the oven to 180°C/350°F/Gas 4. Dust the work surface and a rolling pin with flour and roll out the dough about 1cm/½in thick, dusting the top of it lightly with more flour and occasionally turning it over to stop it sticking. Use a 6cm/2½in diameter pastry cutter to cut out the biscuits, and transfer them to a non-stick baking sheet.

Use a fork to whisk the egg yolk with a little salt, then brush it roughly over the biscuits – you are looking for a patchy effect, not an all-over glaze. Bake for 10–12 minutes, until browned on top and bottom. Heat 1 tbsp honey, either in a small pan or in a bowl in the microwave on Low for 10 seconds. Brush the hot biscuits roughly with the honey, so you get both the golden yolk and the honey showing.

Leave the biscuits to cool slightly and serve them warm, or leave to cool completely and store in an airtight container for up to 3 days.

SMOKY HONEY ALMONDS

Smoked paprika combines with honey to make a tasty bowl of nuts to enjoy with a glass or two before dinner.

It's important to toast the nuts to the right degree. Make sure your oven isn't too hot, or the nuts and honey will burn. Then you need to roll the nuts in the honeyed oil as it cools and thickens so they are evenly coated. Putting the nuts on baking parchment makes life much easier in terms of stickiness not turning to stuckiness.

Whole blanched almonds work best because they show off the colour of the paprika-bronzed oil. This recipe produces a different look and taste to commercial honey-roasted nuts — and is even more addictive.

SERVES 4–6, WITH DRINKS

100G/3½OZ WHOLE BLANCHED ALMONDS

½ TSP SMOKED PAPRIKA

1½ TSP OLIVE OIL

2 TSP HONEY

½ TSP SEA SALT FLAKES

Preheat the oven to 160°C/325°F/Gas 3. Put the nuts on a baking sheet lined with baking parchment and cook for 10 minutes.

Sprinkle the paprika over the nuts, pour over the oil and spoon over the honey (you don't need runny honey — thick honey quickly melts on the hot nuts). Stir around to mix the nuts with the honey, oil and paprika. Put the nuts back in the oven for another 2 minutes, until lightly browned, taking care they do not burn.

Leave the nuts to cool for a couple of minutes. Sprinkle with the salt and use two forks or spoons to stir the nuts around in the honey mixture. Spread the nuts out so they aren't touching each other and leave to cool completely, stirring them once or twice as they cool.

TEATIME BAKING

DRAMBUIE FRUIT CAKE

Drambuie is a Scottish liqueur made with honey and it feels especially appropriate for this fruit cake, although you can use any other sweetish liqueur, such as Cointreau or orange juice. The amount of liquid in the recipe and the hydroscopic properties of honey make for an unashamedly moist cake. I like fruit cake on the damp side and full of goodies: honeyed fruits, ginger, walnuts and citrus peel.

Good to eat straight away, better still after at least a day or two, this cake keeps very well and improves in flavour when stored in an airtight container. A useful cake to make in advance of a holiday so you have it ready for a picnic.

SERVES 12–16

300G/10½OZ UNSALTED BUTTER, AT ROOM
TEMPERATURE, PLUS EXTRA FOR GREASING

300G/10½OZ PLAIN FLOUR, PLUS EXTRA FOR DUSTING

70G/2½OZ LIGHT MUSCOVADO SUGAR

70G/2½OZ CASTER SUGAR

100G/3½OZ HONEY, PLUS 1½–2 TBSP TO GLAZE
(IDEALLY A MEDIUM OR DARK FLAVOURFUL HONEY,
SUCH AS HEATHER OR BUCKWHEAT)

50G/1¾OZ GROUND ALMONDS

PINCH OF SALT

1½ TSP BAKING POWDER

1 TSP MIXED SPICE

½ TBSP FINELY GRATED NUTMEG

5 EGGS

GRATED ZEST OF 1 UNWAXED LEMON

6 PIECES OF PRESERVED STEM GINGER, CHOPPED

100G/3½OZ WALNUT PIECES, ROUGHLY CHOPPED

50G/1¾OZ WHOLE BLANCHED ALMONDS, TO DECORATE

HONEYED FRUITS

150ML/5FL OZ DRAMBUIE OR OTHER LIQUEUR

1 TBSP HONEY

400G/14OZ MIXED DRIED FRUIT (I LIKE TO USE
A MIX WITH CANDIED PEEL)

The night before making the cake (if possible), make the honeyed fruits. Mix the Drambuie with the honey and soak the fruit overnight. If you don't have time, put the fruit and booze in a pan, bring to the boil, turn off the heat and leave to soak for at least 30 minutes.

Preheat the oven to 180°C/350°F/Gas 4. Butter a 26cm/10½in diameter cake tin and lightly dust with flour, discarding the excess.

In a large mixing bowl, cream the butter with the sugars and honey. Put the flour in a bowl with the ground almonds, salt, baking powder, mixed spice and nutmeg. Add the eggs to the creamed mixture one at a time, beating in well and alternating with large spoonfuls of the dry ingredients. Fold in the lemon zest, honeyed fruits, ginger and walnuts. Tip the mixture into the prepared cake tin and smooth the top.

Bake in the centre of the oven for 15 minutes. Take the cake out of the oven and decorate the top with the blanched almonds, pressing the nuts lightly into the surface. Put back in the oven and continue baking for another 45 minutes.

Turn the oven down to 160°C/325°F/Gas 3 for 30 minutes or so, covering the cake with foil to stop the surface from going too brown.

The cake is ready when a skewer inserted into the centre comes out clean. While the cake is still hot, glaze the top by brushing with honey; the honey will melt in the heat. Leave to cool in the tin.

Store in an airtight container for 2–3 weeks (and it keeps for longer than this).

TURKISH PINE NUT, YOGURT AND ORANGE CAKE

A simple technique produces this moist cake that's good for pudding or tea. Everything is mixed together in one bowl, tipped into a cake tin and baked. A syrup of oranges and honey is poured over the cake to give it gloss and extra flavour. That's it. It's the sort of cake you can throw together when half concentrating on something else and it'll still work. I once forgot to add the eggs and the cake still got oohs from the table. And that's ultimately down to the magic ingredient of honey.

MAKES ABOUT 12 SLICES, OR 16 SQUARES

100G/3½OZ PINE NUTS

250G/9OZ GREEK YOGURT

2 LARGE EGGS

250G/9OZ SELF-RAISING FLOUR

1 TSP BAKING POWDER

150ML/5FL OZ SUNFLOWER OIL, PLUS EXTRA FOR GREASING

150G/5½OZ CASTER SUGAR

FINELY GRATED ZEST AND JUICE OF 1 UNWAXED ORANGE

ORANGE SLICES, TO SERVE (OPTIONAL)

ORANGE-HONEY SYRUP AND TOPPING

3 TBSP HONEY, PLUS EXTRA TO GLAZE (A LIGHT HONEY IS BEST AND ORANGE BLOSSOM IS PERFECT)

FINELY GRATED ZEST AND JUICE OF 1 UNWAXED ORANGE

50G/1¾OZ PINE NUTS, TOASTED

Preheat the oven to 180°C/350°F/Gas 4. Toast the pine nuts on a baking sheet in the oven for about 3 minutes, taking great care they do not burn. (You can also toast the nuts in a frying pan over a medium–low heat for 3–4 minutes, stirring frequently, until patched with brown.) Leave the nuts to cool slightly. Line the bottom of a 23cm/9in diameter cake tin (or a 20cm/8in square tin) with baking parchment and grease the sides with a little sunflower oil.

Put all the ingredients for the cake in a large bowl and stir together (or use an electric whisk) until well combined. Pour into the prepared tin and bake for 40–45 minutes, or until a skewer inserted into the centre of the cake comes out clean. Leave to cool in the tin for 10 minutes.

Meanwhile, for the syrup, gently heat the honey with the orange zest and juice in a small pan until the honey has melted. Stir to combine.

Run a knife between the baking parchment and the cake tin to be extra sure that it doesn't stick. Place a wire rack on top of the cake and carefully turn it upside down to turn the paper-lined cake out of the tin. Carefully remove the paper. Put a serving plate on the cake and carefully invert it so the cake is the right-side up on the plate.

Using a skewer, pierce the cake deeply all over. Slowly pour the orange-honey syrup all over the cake so it is absorbed into the dense sponge. Scatter over the pine nuts and lightly brush the top with warmed honey to glaze, rearranging the pine nuts evenly over the cake.

Serve with yogurt, cream or ice cream and, if you like, slices of orange. This can be stored in an airtight container for 3–4 days.

JEWISH HONEY CAKE

Jewish traditions use honey to celebrate events in a richly symbolic way. Honey is eaten on a child's first day at school, showing the sweetness of knowledge, and a sweet, spicy honey cake is on the table for Rosh Hashanah (Jewish New Year) to wish good fortune and happiness for the months ahead; what could be happier than honey?

A friend, Rosalind Jerram, always makes honey cake to wish family and friends a happy future year. What makes it a special tradition? 'It's something about the sweetness of the honey that I feel will nurture for the year ahead, and also that I imagine women of past generations in my family making this every Rosh Hashanah,' she explains. When Rosalind's daughter Daphne had just been born, even in the midst of new-baby exhaustion, it felt especially important to make the cake to welcome Daphne and the new year ahead.

Rosalind's recipe came originally from A Treasury of Jewish Holiday Baking *by Marcy Goldman and this is an adapted version. There are many slight variations of Jewish honey cake. I've used cinnamon, cloves and allspice, but you can vary the quantities, or the spices, according to taste, perhaps adding freshly grated nutmeg and ground ginger. I've made this as a traybake whereas for Rosh Hashanah it might be baked in a special tin, such as a bundt tin, to give it more impact.*

MAKES ABOUT 30 SQUARES

200G/7OZ HONEY (LIGHT OR MEDIUM)

100G/3½OZ GRANULATED SUGAR

100G/3½OZ SOFT BROWN SUGAR

150ML/5FL OZ VEGETABLE OIL

100ML/3½FL OZ STRONG COFFEE

100ML/3½FL OZ ORANGE JUICE (JUICE OF ABOUT 1½ ORANGES)

FINELY GRATED ZEST OF 1 UNWAXED ORANGE

2 TBSP RUM, BRANDY OR BOURBON WHISKY

300G/10½OZ PLAIN FLOUR

1½ TSP BAKING POWDER

1 TSP BICARBONATE OF SODA

¼ TSP SEA SALT

2 TSP GROUND CINNAMON

¼ TSP GROUND CLOVES

¼ TSP GROUND ALLSPICE

2 LARGE EGGS

1 TSP VANILLA EXTRACT

Preheat the oven to 180°C/350°F/Gas 4. Line a baking tin (30 x 22 x 4cm/12 x 8½ x 1½in) with baking parchment.

Put the honey, sugars, oil, coffee, orange juice, orange zest and rum or other spirit in a small pan. Heat gently, stirring occasionally, until the sugar has dissolved. Leave to cool slightly.

Meanwhile, put the flour, baking powder, bicarbonate of soda and salt into a large mixing bowl. Add the spices and stir everything together.

Stir the eggs and vanilla into the honey mixture, then tip the wet mixture into the dry ingredients. Use a hand-held electric whisk or a wooden spoon to beat all the ingredients together thoroughly.

Scrape the mixture into the prepared tin. Bake for about 30 minutes, or until the top is glossy and springy and starting to brown and a skewer inserted into the centre comes out clean (a cake made in a more upright tin will take 10 minutes or so longer). Leave to cool in the tin, then wrap in greaseproof paper and clingfilm and store in an airtight container. This keeps well and even improves and mellows with time. Ideally, keep it for a couple of days before eating, but it is also good cut into straight away.

HAZELNUT AND HONEY SPONGE

Here's a light sponge that is all about honey and nuts. Different honeys each have their own flavour characteristics and you can certainly taste them in this cake, especially in the glaze. You could use a less expensive honey in the cake itself and save a special one for the top. As for the nuts, toasting, skinning and chopping the hazelnuts does take time, but it will give you the very best texture and taste. This cake is good enough to be served for pudding with poached or fresh fruit and lightly whipped cream, or as a teatime treat. The nuts and honey also make it a nourishing slice at odd moments of the day.

SERVES 8–12

BUTTER FOR GREASING

100G/3½OZ PLAIN FLOUR, PLUS 1 TBSP FOR DUSTING

300G/10½OZ HAZELNUTS, SKIN ON

8 LARGE EGGS, SEPARATED

100G/3½OZ HONEY (A MEDIUM OR DARK ONE SUCH AS GREEK OR HEATHER IS BEST)

50G/1¾OZ CASTER SUGAR

FINELY GRATED ZEST OF 1 UNWAXED LEMON

½ TSP SEA SALT

GLAZE

2 TBSP HONEY

JUICE OF 1 LEMON

Preheat the oven to 160°C/325°F/Gas 3. Butter the bottom and sides of a 23cm/9in diameter cake tin (a springform tin is best). Sprinkle 1 tbsp flour into the tin and shake it around so the flour lightly coats the bottom and sides, then tip out any excess, or line the tin with baking parchment.

Put the hazelnuts in a single layer on a baking sheet. Toast the nuts in the oven for 15 minutes, turning the baking sheet around once to ensure they brown evenly and watching that they don't burn. Leave until cool enough to handle. Put the nuts onto a clean, dry tea towel, fold over and rub them to remove most of their skin. Roughly pick off any remaining large bits of skin – there's no need to be perfectionist about this. Pulse the nuts in a food processor to chop them medium–fine.

Turn the oven up to 180°C/350°F/Gas 4. Put the egg yolks in a mixing bowl with the honey and sugar and whisk for about 5 minutes, until creamy and thick. Whisk in the lemon zest, then use a big metal spoon to fold in the flour, salt and chopped nuts.

Whisk the egg whites until stiff. Stir about a quarter of the yolk mixture into the whites, then lightly fold in the rest so it is thoroughly mixed, keeping as much air in the mixture as possible. Tip the mixture into the prepared tin and bake for 15 minutes. Turn the heat down to 170°C/340°F/Gas 3½ and continue baking for 20–25 minutes, until the top is brown and a skewer inserted into the centre of the cake comes out clean. Leave to cool in the tin.

Place the cake on a serving plate. For the glaze, gently warm the honey in the microwave on Low for 10 seconds, or warm briefly in a saucepan, and mix with the lemon juice. Brush this glaze all over the cake. Store in an airtight container for up to a week.

HONEY AND LEMON DRIZZLE CAKE

Drizzle cake is even more delicious when you scent the sponge with fragrant honey. I use sugar to make the sponge and honey in the syrup, heating it very gently so none of the flavour is lost.

SERVES 8–10

125G/4½OZ UNSALTED BUTTER, AT ROOM TEMPERATURE, PLUS EXTRA FOR GREASING

125G/4½OZ CASTER SUGAR (OR 75G/2¾OZ CASTER AND 50G/1¾OZ SOFT BROWN SUGAR)

PINCH OF SALT

2 LARGE EGGS

225G/8OZ SELF-RAISING FLOUR

3 CARDAMOM PODS

FINELY GRATED ZEST AND JUICE OF 1 UNWAXED LEMON

HONEY-LEMON DRIZZLE

JUICE OF 1½ LEMONS

60G/2¼OZ HONEY (A LIGHT OR MEDIUM ONE SUCH AS ORANGE BLOSSOM OR LAVENDER)

Preheat the oven to 190°C/375°F/Gas 5. Butter a 900g/2lb loaf tin (about 25 x 12 x 6cm/10 x 4½ x 2½in).

Put the butter in a food processor and whizz. Add the sugar and salt and whizz to mix. Add one egg and whizz again, then add half the flour, then the second egg and finally the rest of the flour, pulsing briefly to mix after each addition. (Alternatively, beat the mixture by hand using a wooden spoon.)

Use a small sharp knife to slit open the cardamom pods. Remove the seeds and crush with a pestle and mortar. Add to the cake mixture with the lemon zest and juice. Pulse the food processor to mix, or beat in with a wooden spoon.

Spoon the mixture into the prepared tin and bake for 45 minutes, or until the mixture is risen, golden and a skewer inserted into the centre comes out clean. Leave to cool in the tin for 10 minutes.

Meanwhile, put the lemon juice and honey for the drizzle in a small pan. Heat gently to melt the honey and mix the two together.

Take the cake out of the tin and put it on a plate. Using a skewer, pierce the top of the cake deeply. Carefully and slowly pour the drizzle mixture all over the top of the cake. Store in an airtight container. This is best eaten within 3 or 4 days.

HONEY BANANA BREAD

Banana bread is one of those foods that satisfies a gap and gives you some honey-zoom at breakfast, lunch, tea or supper — and all points in between.

One of the easiest of cakes to make, banana bread benefits greatly from the added richness of honey. I've used fudgy muscovado sugar as well, to complement the honey's sweetness and flavour. If you have plain brown flour, it will add a bit more substance and nutrition, but all white flour is fine.

MAKES 1 LARGE LOAF

BUTTER FOR GREASING (OPTIONAL)

100G/3½OZ SULTANAS

2 TBSP RUM AND 2 TBSP WATER, OR 4 TBSP FRUIT JUICE

100G/3½OZ PLAIN WHITE FLOUR

75G/2¾OZ PLAIN BROWN FLOUR (OR USE WHITE)

2 TSP BAKING POWDER

½ TSP BICARBONATE OF SODA

LARGE PINCH OF SALT

3 RIPE SMALL BANANAS

150G/5½OZ NATURAL YOGURT

1 TSP VANILLA EXTRACT

100G/3½OZ LIGHT MUSCOVADO SUGAR

50G/1¾OZ HONEY, PLUS ABOUT 1 TBSP TO GLAZE

2 LARGE EGGS

100G/3½OZ BRAZIL NUTS, ROUGHLY CHOPPED

Preheat the oven to 170°C/340°F/Gas 3½. Butter a large loaf tin (23 x 13 x 7cm/9 x 5 x 2¾in) or line it with baking parchment.

Put the sultanas in a small pan with the rum and water or fruit juice. Bring to the boil and immediately turn off the heat. Leave the fruit to soak up the liquid while you prepare the other ingredients.

Put the flours, baking powder, bicarbonate of soda and salt in a large mixing bowl and set aside.

Put the bananas in a small bowl and mash them with a fork. Stir in the yogurt and vanilla. Put the sugar and honey into another bowl and crack in the eggs. Using an electric whisk, mix the sugar, honey and eggs together until they have lightened in colour and thickened (about 5 minutes). Stir in the banana and yogurt mixture.

Fold in the dry ingredients, the soaked sultanas and three quarters of the chopped nuts.

Gently scrape the mixture into the prepared tin. Scatter the remaining nuts over the top. Bake for about 1 hour, or until a skewer inserted into the centre comes out clean.

Gently warm a little honey in the microwave on Low for 10 seconds, or warm briefly in a saucepan. Brush it over the banana bread. Leave the banana bread in the tin to cool. Wrap in greaseproof paper and store in an airtight container for a week or so.

HONEY NUT BLONDIES

Blondies are the even more indulgent white chocolate version of brownies. These use honey as one of the sweeteners, together with sugar and the white chocolate itself. Dried cranberries or cherries offset the sweetness and the nuts add some texture and heft, along with extra chunks of chocolate.

These dense little nuggets of honeyed sweetness are great with a cup of coffee. I also love them as a pudding, served with raspberries, which are a great pairing with both honey and white chocolate.

MAKES 16 SQUARES

100G/3½OZ UNSALTED BUTTER, PLUS EXTRA
FOR GREASING

175G/6OZ WHITE CHOCOLATE

100G/3½OZ HONEY (LIGHT IS BEST)

50G/1¾OZ LIGHT MUSCOVADO SUGAR

2 EGGS, LIGHTLY BEATEN

1½ TSP VANILLA EXTRACT

200G/7OZ PLAIN WHITE FLOUR

1½ TSP BAKING POWDER

100G/3½OZ MACADAMIA OR BRAZIL NUTS, CHOPPED

50G/1¾OZ DRIED CHERRIES OR CRANBERRIES

Preheat the oven to 170°C/340°F/Gas 3½. Butter a 20cm/8in square shallow baking tin. Chop 75g/2¾oz of the white chocolate into medium–small chunks. Break the rest of the chocolate into squares.

Melt the butter and the squares of chocolate together in a heatproof bowl over a pan of simmering water. White chocolate has a tendency to split because of its high fat content, so be careful not to overheat. As the chocolate and butter melt, stir them together to amalgamate.

Remove the bowl from the heat and stir in the honey and sugar. Next stir in the eggs and vanilla. Fold in the flour and baking powder, then the nuts, dried fruit and white chocolate chunks.

Scrape this stiff dough into the prepared tin and bake for 20–25 minutes, until the top is lightly brown all over, and slightly darker at the sides. The inside will still be a bit moist when you insert a skewer into the centre. Leave in the tin to cool completely. Cut into 16 squares and store in an airtight container. These are best eaten within a week.

MADELEINES

Some recipes really show 'value for honey'. These classic little buttery cakes, traditionally made in shell-shaped moulds, use a small amount of honey to great effect. It is worth dipping into one of the more aromatic honeys, such as heather or thyme, to get a flavour that scents the sponge beautifully.

Over the years, I seem to have built up boxfuls of baking paraphernalia. Not all of these items are worth their space, but a madeleine mould is an object of beauty. I could use another bun tin but have never regretted having the perfect one.

Madeleines are great for a posh tea, perhaps with an elegant cup of Darjeeling, or with fruit salad for a pudding.

MAKES 20—24

100G/3½OZ UNSALTED BUTTER, PLUS EXTRA FOR GREASING

2 TBSP HONEY (A FRAGRANT MEDIUM OR DARK HONEY SUCH AS HEATHER IS BEST)

3 EGGS

100G/3½OZ CASTER SUGAR

100G/3½OZ SELF-RAISING FLOUR

50G/1¾OZ GROUND ALMONDS

PINCH OF SEA SALT

Melt the butter and honey together in a small pan over a low heat. Leave to cool.

Put the eggs in a large bowl, add the sugar and mix together using an electric whisk for 10 minutes, until tripled in volume. Whisk in the cooled honey butter.

Carefully fold in the flour, almonds and salt until the mixture is thoroughly combined but holding as much air as possible.

Ideally, leave the mixture to rest, covered, in the fridge for 2 hours. This makes it easier to spoon into the moulds and also means it rises slightly more and holds its shape.

Preheat the oven to 180°C/350°F/Gas 4. Grease a madeleine mould or a shallow bun tin with butter. Put 2 teaspoons of the mixture in each mould, but don't overfill. Bake for 7 minutes or so, until light brown; check the cakes after 5 minutes, as honey burns quickly, and remember that the madeleines will be browner on the bottom than on the top.

Leave to cool in the mould for 5—10 minutes, and then carefully unmould the madeleines and place on a wire rack to cool completely. Store in an airtight container and eat within a couple of days.

LUCKY LEBKUCHEN PIGS

German-speaking countries have a long-standing tradition, dating back to at least the 14th century, of honey-and-spice cakes and biscuits that are made to celebrate feast days and holidays. Some of these are made into shapes, including pigs, which are regarded as creatures of good fortune and prosperity. This recipe is based on one in Festive Baking in Austria, Germany and Switzerland *by Sarah Kelly, an excellent book that includes many different kinds of honey cakes and biscuits.*

You can make the dough up to 3 days in advance: wrap it in greaseproof paper and store at room temperature. You will get slightly less puffy pigs if you don't use the dough immediately. You can buy a special pig-shaped cutter (about 8 x 5cm/3¼ x 2in) online, or you can make a template, cutting out a pig shape from a stiffish piece of cardboard.

MAKE ABOUT 35 PIGS
325G/11½OZ PLAIN FLOUR, PLUS EXTRA FOR DUSTING
½ TSP BICARBONATE OF SODA
100G/3½OZ HONEY (A MEDIUM OR DARK ONE WORKS WELL)
100G/3½OZ CASTER SUGAR
70G/2½OZ UNSALTED BUTTER
1 EGG, LIGHTLY BEATEN
WARMED HONEY, TO GLAZE
GLACÉ ICING, TO DECORATE (OPTIONAL)

SPICE MIX
¾ TSP GROUND CINNAMON
½ TSP GROUND CARDAMOM
¼ TSP GROUND CLOVES
¼ TSP GROUND ANISE
½ TSP GRATED NUTMEG
1 TSP GROUND GINGER

Put the flour in a bowl and add the bicarbonate of soda. Heat the honey, sugar and butter together in a saucepan over a low heat until the butter has melted and the sugar dissolved. Don't bring to the boil.

Stir the spices into the saucepan, then gradually beat in the flour, adding as much as you need to make a dough that pulls away from sides of pan as you stir. Stir in the beaten egg.

Tip the mixture onto a floured surface and knead until smooth, adding more flour if necessary to get a nicely malleable dough.

Preheat the oven to 180°C/350°F/Gas 4. Cut the dough into quarters to make it easier to handle. Roll out each piece to ½–1cm/about ⅜ in thick. Cut out pig shapes, using a pig-shaped cutter or cardboard template.

Put the pigs on a baking sheet, leaving about 2cm/¾in between each pig. Brush with warmed honey and bake for 8 minutes, or until lightly browned, turning the baking sheet around once during cooking if necessary so they cook evenly. Leave the pigs on the tin to firm up for a few minutes, then transfer to a wire rack to cool.

When cool, you can decorate them with glacé icing (you can buy this ready-made in little piping tubes or put some icing sugar in a small bowl and add just enough just-boiled water to make a smooth, glossy, but flowing icing). If you pipe the initials or names of your friends, the pigs make a charming gift or table place setting.

These keep for months in an airtight container. If they get a bit hard, the traditional German trick to soften them is to leave a quarter apple in the container for a few hours.

Hattie

POLLEN BUTTER SHORTBREAD

Shortbread is so simple and so good. There are several secrets for perfection. Firstly, the butter matters. French butter has a higher butterfat content than British and its slightly firmer character really does make a difference. Granulated sugar gives a better texture than caster sugar, a tip from food writer Rose Prince. A high proportion of fine semolina to plain flour makes this version extra-short, or crumbly — you press the dough into the tin rather than rolling it out. For the best result, leave the dough to chill before cooking it.

After experimenting, this recipe contains no honey: I use sugar because the way the crystals melt apparently creates minuscule holes in the dough — yet another part of the delectable texture. I've included the recipe in this book because of the dusting of pollen: the golden flecks combine with the snowy sugar to make this a shortbread with a difference.

MAKES 8 TRIANGLES
100G/3½OZ PLAIN FLOUR
100G/3½OZ FINE SEMOLINA
70G/2½OZ GRANULATED SUGAR, PLUS 2 TSP FOR TOP
125G/4½OZ BEST-QUALITY BUTTER (SALTED OR UNSALTED), FRIDGE COLD, CUT INTO SMALL PIECES
LARGE PINCH OF SEA SALT (IF BUTTER IS UNSALTED)
1½ TSP POLLEN

Mix the flour and semolina together in a mixing bowl and stir in the sugar. Rub the butter into the dry ingredients until the mixture resembles breadcrumbs. Bring together with your hands to form a rough, slightly crumbly ball of dough.

Press the dough into a 20cm/8in diameter tin so it is about 2cm/¾in thick. Cover with clingfilm and put in the fridge for at least 1 hour and ideally overnight.

Preheat the oven to 150°C/300°F/Gas 2. Prick the shortbread all over to ensure it cooks through. Put the tin in the centre of the oven and bake for about 40–45 minutes, checking after 30 minutes and turning the tin around a couple of times during cooking to ensure it bakes evenly. The shortbread is ready when light brown on top.

Remove from the oven and leave for 2 minutes. Cut down into the shortbread to make eight triangles (but don't pull them apart yet). Scatter 2 tsp sugar over the shortbread, then scatter the pollen over the top and push the grains down lightly with the back of the spoon to help them stick.

Leave the shortbread in the tin to cool completely, then break into triangles. Store in an airtight container for up to a week.

HONEY OAT COOKIES

Honey attracts moisture and so instead of a crisp biscuit this is a cookie with a rich, moist texture. The oats make them a nourishing snack and they show off the taste of honey. I like to keep them simple, but you could add cinnamon or other spices and also currants or other dried fruit, chopped nuts or chopped chocolate.

MAKES 24

125G/4½OZ UNSALTED BUTTER, AT ROOM TEMPERATURE

50G/1¾OZ SOFT BROWN SUGAR

100G/3½OZ HONEY

1 TSP VANILLA EXTRACT

4 TBSP MILK

1 EGG

100G/3½OZ PLAIN FLOUR

150G/5½OZ ROLLED PORRIDGE OATS (QUICK-COOK, NOT JUMBO)

1 TSP BICARBONATE OF SODA

PINCH OF SALT

BIG PINCH OF GROUND CINNAMON (OPTIONAL)

Preheat the oven to 180°C/350°F/Gas 4. Line two baking sheets with baking parchment or grease them well with butter.

Put the butter, sugar and honey in a large mixing bowl and cream together until pale and fluffy. Mix in the vanilla, milk and egg. Then mix in the flour, oats, bicarbonate of soda and salt, and the cinnamon, if using.

Drop tablespoonfuls of the cookie dough onto the baking sheets and press down with the back of the spoon to flatten slightly into thickish rough rounds. These cookies do not spread, so there is no need to leave much space between them.

Cook for 13–14 minutes, until brown all over, turning the baking sheets around in the oven halfway through, if necessary, to ensure the biscuits cook evenly. Leave to cool slightly on the baking sheet, then transfer to a wire rack to cool completely. Store in an airtight container. You can eat these straight away, but the honey flavour develops and they are best eaten the day after cooking.

PAIN D'ÉPICES

Pain d'épices is a French spice bread that was traditionally made by fermenting the mixture over some weeks, rather like a sourdough bread. Nowadays most home bakers make it as a cake. It originally used rye flour, and it is great to use rye flour alongside white, though just white is fine. You can use all honey, but I mix the honey with dark sugar. This is a dark cake and a dark or medium honey goes well with the spices.

Vary the spice mix according to your taste. I keep a careful eye on the ground cloves but get more liberal with the nutmeg. There isn't too much ginger, either, so it remains a spice cake rather than a gingerbread. You could also experiment with coriander, a flavour that goes well with citrus and honey.

Pain d'épices is great for teatime, but its spicy nature means it also goes well with cheese, or try it lightly toasted with chicken liver pâté.

MAKES 1 LARGE LOAF

150G/5½OZ UNSALTED BUTTER, PLUS EXTRA
FOR GREASING

150G/5½OZ HONEY (A DARK ONE IS BEST,
SUCH AS GREEK MOUNTAIN HONEY OR BUCKWHEAT),
PLUS EXTRA TO GLAZE (OPTIONAL)

100G/3½OZ WHITE FLOUR

100G/3½OZ RYE FLOUR

LARGE PINCH OF SEA SALT

1 TSP BICARBONATE OF SODA

100G/3½OZ DARK MUSCOVADO SUGAR

1 TSP VANILLA EXTRACT

FINELY GRATED ZEST OF 1 UNWAXED ORANGE

FINELY GRATED ZEST OF 1 UNWAXED LEMON

3 EGGS

150ML/5FL OZ SEMI-SKIMMED MILK

SPICE MIX

½ TSP GROUND NUTMEG

½ TSP GROUND GINGER

½ TSP GROUND CINNAMON

¼ TSP GROUND CLOVES

¼ TSP GROUND ANISE

Preheat the oven to 160°C/325°F/Gas 3. Butter a non-stick loaf tin (23 x 13 x 7cm/9 x 5 x 2¾in).

Put the butter and honey in a small pan and gently heat, stirring occasionally, until the butter has melted into the honey. Leave to one side.

Put the flours, salt and bicarbonate of soda into a large bowl. Stir in the sugar and the spice mix. Stir in the honey and butter mixture with the vanilla and citrus zest. Add the eggs and milk.

Pour the mixture into the prepared tin and bake for around 1 hour and 10 minutes, or until a skewer inserted into the centre of the cake comes out clean. Check after 1 hour and then every 5 minutes or so after that.

Leave to cool in the tin, then turn out and, if you like, brush with honey, warmed through to make it more runny. Store in an airtight container. This will keep for several weeks.

DITTY'S HONEY-FRUITED SODA BREAD

Robert Ditty runs a wonderful bakery, Ditty's, in Castledawson, a 40-minute drive from Belfast in Northern Ireland. He is also a beekeeper, with 26 hives in four apiaries. Two of his beekeeping sites are in the parkland of a large house nearby. The many unusual shrubs and trees here make what Robert calls a 'super-duper foodhall for bees'.

Some of Robert's bees make the most of the pink Himalayan balsam that grows profusely on the riverbanks. Considered a weed by many, balsam is a bonus for the bees and Robert credits some of his honey prizes to the plant's nectar. He knows when the bees have been on these blooms because they are covered in grey pollen that, he says, looks almost like flour.

Another of Robert's apiary sites is beside his home, where he has an orchard, and he also encourages nearby farmers to grow clover to feed the bees as well as to fix nitrogen in their soil (growing clover is one of the principles of green agriculture). More recently he put a hive near the bakery, now that people are getting more accustomed to the idea of urban beekeeping. Customers certainly like the honey, which sells out almost the moment it reaches the shop.

Honey and flour are good companions in breadmaking. Honey softens the crumb and adds to the keeping quality of breads, says Robert. He makes honey cookies, a honey and spelt bread and honey ice cream for the bakery's excellent café. This is a simple soda bread using honey-flavoured dried fruits. How to eat it? 'Hot. With butter,' says Robert. And honey, of course.

SERVES 6

40G/1½OZ HONEY

150G/5½OZ MIXED DRIED FRUIT

300G/10½OZ SELF-RAISING FLOUR, PLUS EXTRA FOR DUSTING

LARGE PINCH OF FINE SEA SALT

200ML/7FL OZ BUTTERMILK

2 TBSP OLIVE OIL

At least 30 minutes before making the bread, or the night before, mix the honey with 3 tbsp tepid water, stirring so it dissolves. Stir in the dried fruit and rub so they are well coated in the honey syrup. Leave to soak.

Preheat the oven to 220°C/425°F/Gas 7. Sift the flour into a large bowl and add the salt. Add the honeyed dried fruits and mix well, using a wooden spoon.

Make a well in the centre of the flour mixture. Pour the buttermilk and oil into the well and mix with a wooden spoon or one hand until the ingredients come together.

Transfer the dough to a floured surface. Dust the top with flour and knead thoroughly until the dough feels even, then shape into a rough round. Place on a baking sheet and press out to form a circle about 20cm/8in in diameter. Mark a cross shape into the dough with a knife, going almost down to the baking sheet.

Bake for 25–30 minutes, or until the crust is brown all over. Best eaten on the same day, or keep in an airtight container for a day or two and serve toasted.

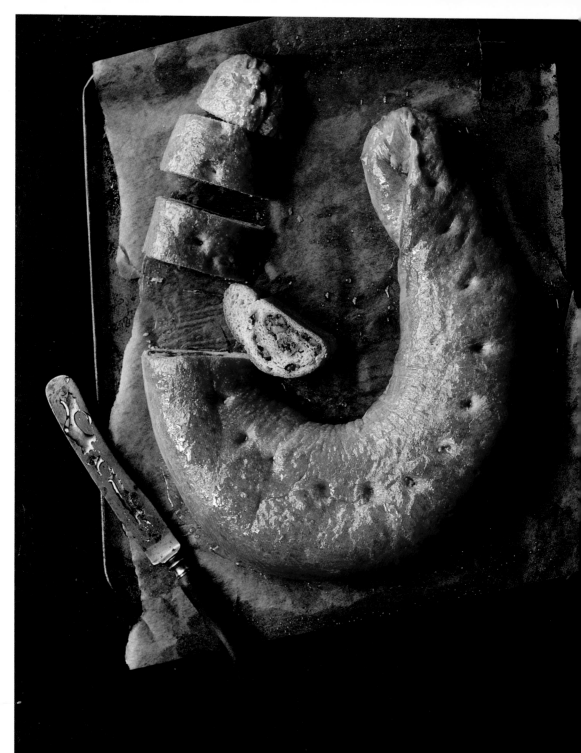

HUNGARIAN WALNUT HORSESHOE

This recipe is based on one from a wonderful collection of honey recipes from around the world, Honey, I'm Homemade, *gathered by May Berenbaum, an American entomologist. Bee-people seem to associate like their subjects, passing on stories and recipes like scout bees coming back to the hive with the scent of a good nectar flow. This attractive horseshoe loaf came from a fellow entomologist who supplemented his studies by teaching English to students in Hungary and was paid in part by treats like this from students' grandmothers. One of the special aspects of this enriched bread is the way the honey in the filling permeates the dough, especially if left overnight, and combines with the walnuts to make a quietly nourishing whole. It's good with mid-morning coffee, afternoon tea, or for breakfast or brunch with a small slice of cheese on top.*

MAKES 1 LOAF

250G/9OZ STRONG WHITE BREAD FLOUR

1½ TSP CASTER SUGAR

6G EASY-BLEND YEAST (MOST OF A 7G PACKET)

¼ TSP FINE SEA SALT

125ML/4FL OZ FULL-FAT MILK

100G/3½OZ UNSALTED BUTTER, AT ROOM TEMPERATURE, CUT INTO PIECES

½–1 WHOLE EGG, BEATEN, TO GLAZE

1–2 TBSP RUNNY HONEY, TO GLAZE (OPTIONAL)

WALNUT-HONEY FILLING

60G/2¼OZ HONEY (THIS CAN TAKE A MEDIUM OR DARK HONEY SUCH AS THYME OR CHESTNUT)

2 TBSP MILK

½ TSP VANILLA EXTRACT

150G/5½OZ WALNUT PIECES, CHOPPED SMALL

2 TBSP FRESH WHITE BREADCRUMBS

Mix together the flour, sugar, yeast and salt in a large bowl. Make a well in the centre and pour in the milk. Use the fingers of one hand to draw the dry ingredients into the liquid, adding more milk if necessary, until the flour comes away from the side of the bowl.

Bring the dough together and knead on a lightly floured or oiled work surface for 7 minutes, or in a mixer for 5 minutes. Add the butter, a few pieces at a time, kneading as you go. Knead for another 5 minutes, or 3 minutes in a mixer.

Roll the dough into a ball and put it in a lightly greased bowl, then cover and leave in a warmish place for 1–2 hours, or until doubled in size (as the dough is enriched with butter it can take longer to rise than standard dough).

Meanwhile, make the filling: melt the honey in a small pan, stir in the milk and vanilla, then the walnuts and breadcrumbs.

Knock back the dough and roll it out into a long, thin rectangle. Spread the filling over the rectangle and roll up into a long sausage, then bend around into a horseshoe shape. Press the dough at each end of the horseshoe to seal in the filling and make neat ends.

Line a baking sheet with baking parchment, carefully place the horseshoe on top and put the baking sheet in a large bag that doesn't touch the top of the dough. Leave to rise for 1 hour, or until doubled in size.

Meanwhile, preheat the oven to 200°C/400°F/Gas 6. Whisk the egg and use as much as you can to lacquer the outside of the bread. Gently poke holes in the horseshoe as if for nails.

Bake for 25 minutes, or until well risen and brown. Leave to cool on a wire rack. If you like, glaze with honey: gently warm the honey in the microwave on Low for 10 seconds, or warm briefly in a saucepan.

To serve, cut into thick slices. Store in an airtight container and eat within a couple of days.

PUDDINGS

ELDERFLOWER FRITTERS

The subtly scented muskiness of elderflowers comes through in each mouthful of these crisp fritters. I've added a touch of elderflower cordial to the batter to strengthen the flavour, but if you pick fresh flowers on a sunny day, then there should already be plenty. A drizzle of light honey enhances the floral character of the dish. This is a quick dish to make and a good one for children: they can pick the flowers and then drizzle on the honey and enjoy the crunchy sweet fritters as soon as they are cool enough to eat.

SERVES 6

6 LARGE, FRESH ELDERFLOWER HEADS

115G/4OZ PLAIN FLOUR

SMALL PINCH OF SALT

175ML/6FL OZ WATER

1½ TBSP ELDERFLOWER CORDIAL

1 LITRE/1¾ PINTS VEGETABLE OIL

1 EGG WHITE

TO FINISH

ABOUT 6 TBSP HONEY (LIGHT IS BEST, SUCH AS ACACIA OR ORANGE BLOSSOM)

LEMON WEDGES

It is best to pick the elderflowers on the day you want to cook them and make sure they are fresh and fragrant. Pick over the elderflowers, removing any small insects or browning flowers. Divide each head into florets, leaving a bit of stalk on each for the eater to hold.

Put the flour and salt in a mixing bowl. Pour the water and cordial into a measuring jug and whisk together with a fork. Pour the liquid into the centre of the flour and gradually whisk in the flour, drawing it into the liquid until you have a batter with the consistency of double cream. Leave in a cool place for 30 minutes.

Heat the oil in a large deep pan until a small piece of batter dropped into it fizzles immediately and rises to the surface (about 190°C/375°F). While the oil heats up, whisk the egg white until stiff and gently fold into the batter.

Dip the elderflower florets into the batter and then lower into the hot oil, two or three at a time so you do not overcrowd the pan. You don't need to turn them over. When very lightly brown, remove with a slotted spoon and drain on kitchen paper while you get on with the next batch. These are best eaten straight away, drizzled with honey — about 1 tsp per floret — and a squeeze of lemon juice.

POLLINATORS' FRUIT SALAD

This is a celebration of some of the fruits and nuts that are pollinated by bees. Many kinds of plant food rely on insect pollination and honey bees are numerous and important pollinators. This role is so crucial that honey can be considered as a by-product of a much more important function. Without bees, we'd have staples to eat such as bread and butter but not the jam on top; the porridge, not the berries. We'd survive; but without glory or joy. So here's to bees and fruition!

Mangoes, raspberries and pistachios all need bees for pollination. I've seen the cascading mass of pinkish flowers that make up mango blossom – and could hardly believe that the tiny buds will become large pendulous fruits full of sweet flesh and juice. What's even more amazing is that only a few buds on each floret become fruit even though many are pollinated. 'The tree knows what it can do,' said the agronomist on the farm that I was visiting in north-east Brazil. I came away with a bottle of fruity mango honey that I've used happily in this salad, although many different kinds of honey work well here.

I like to keep fruit salads simple and use just one or two kinds of fruit. The syrup in this fruit salad is inspired by the mixture of wine and honey that was drunk by the ancient Romans. The syrup is very much part of the dish and not just a dressing for the fruit. This makes a wonderful brunch dish. I serve it without cream, but others may feel a fruit salad is naked without it.

SERVES 6

200ML/7FL OZ WHITE WINE

2 TBSP HONEY (PREFERABLY A LIGHT ONE SUCH AS ACACIA OR ORANGE BLOSSOM, OR A MEDIUM ONE SUCH AS THYME, OR EVEN MANGO, IF YOU CAN FIND IT)

2 LARGE OR 3 MEDIUM MANGOES

250G/9OZ RASPBERRIES (THAWED FROZEN ARE FINE)

50G/1¾OZ PISTACHIO NUTS

Mix together the wine and honey in a serving bowl. Let the honey dissolve, stirring occasionally, while you prepare the fruit.

Slice the 'cheeks' off the mangoes and score the flesh down to the skin into roughly 3cm/1¼in cubes. Invert the cheek so the cubes stand out and cut them away from the skin. Cut the skin off the sides of the mango and cut the flesh off the stone into small chunks.

Add the mango pieces to the honeyed wine, along with any juices. Carefully stir in the raspberries. Cover and chill for an hour or more. Just before serving, roughly chop the pistachios and scatter over the top.

This is best eaten within a few hours but is also fine the next day.

GOOSEBERRY (OR RHUBARB), HONEY AND SAFFRON FOOL

Honey and saffron are golden partners and work brilliantly to sweeten and enhance tart fruit such as gooseberries or rhubarb. The fruit's sharpness is further softened with cream and custard to make a two-layered fool with a silky fruit-custard base and an ethereal honey-saffron cream top.

SERVES 6–8

¼ TSP SAFFRON STRANDS

300ML/10FL OZ WHIPPING CREAM

ABOUT 2 TBSP HONEY (A LIGHT ONE SUCH AS WILDFLOWER, ACACIA OR ORANGE BLOSSOM OR A FRAGRANT MEDIUM ONE SUCH AS HEATHER)

300ML/10FL OZ READY-MADE CUSTARD

A HANDFUL OF TOASTED FLAKED ALMONDS (OPTIONAL)

HONEYED GOOSEBERRIES OR RHUBARB

250G/9OZ GOOSEBERRIES, TOPPED AND TAILED, OR 250G/9OZ RHUBARB, CUT INTO 2CM/¾IN SLICES

2 TBSP HONEY

2 TBSP ELDERFLOWER CORDIAL, OR 1½ TBSP SOFT BROWN SUGAR

Crush the saffron between your fingers and put in a small bowl with 2 tsp cold water. Leave to soak for 10 minutes.

For the honeyed gooseberries or rhubarb, put the fruit in a small pan with the honey and cordial or sugar and 2 tbsp water. Cover with a lid and cook over a medium heat for 4–5 minutes, or until the fruit softens to the point of collapse, stirring occasionally. Mash the fruit into a rough purée using a wooden spoon. Leave to cool completely.

Whip the cream until soft peaks form. Whisk in the honey, tasting to ensure you get the right amount of sweetness – it will depend on your palate, the fruit and the type of honey.

Pour the saffron water into the cream; put the strands in a pestle and mortar and crush, then add to the cream. Whisk together, removing any saffron strands from the whisk as you go.

In another bowl, stir together the fruit purée and the custard. Spoon the fruit-custard fool into 6–8 glasses or small bowls, and spoon the saffron cream on top. Chill for at least 30 minutes and up to 24 hours – if leaving for more than 1 hour, cover the fools with clingfilm. If you like, scatter over some toasted flaked almonds just before serving.

GREEK RICOTTA AND HONEY PIE

This recipe is adapted from the wonderful Honey, I'm Homemade *(see the introduction to Hungarian walnut horseshoe, page 131). Called* melopita *in Greece, the recipe comes originally from the island of Sifnos in the Cyclades. I've been to this beautiful island, famous for its honey and the mellifluous words of poets, and was drawn to this tasty, rustic cheesecake with plenty of honey flavour.*

The honey from Sifnos was considered to be one of the finest in ancient Greece and remains so today. Vivid blue hives stand on the steep slopes of the island above the blue sea. My friend Debs gave me some honey from Sifnos one year; when I opened the lid and inhaled I was transported to the herb-scented hillside of this Greek island.

This is delicious on its own or served with fruit; apricots work well, adding a slight acidity to counterbalance the honey-sweet pie.

SERVES 10—12

500G/1LB 2OZ RICOTTA

2 EGGS, LIGHTLY BEATEN

100G/3½OZ CASTER SUGAR

150G/5½OZ HONEY (LIGHT OR MEDIUM; GREEK HONEY
FOR PREFERENCE)

3 TBSP DOUBLE CREAM

1 TSP FINELY GRATED ORANGE OR LEMON ZEST
(USE UNWAXED FRUIT)

¼ TSP GRATED NUTMEG

¼ TSP CINNAMON

1½ TBSP TOASTED SESAME SEEDS

PASTRY

2 TBSP TOASTED SESAME SEEDS

115G/4OZ PLAIN FLOUR

60G/2¼OZ COLD UNSALTED BUTTER, CUT INTO SMALL CUBES

1 TBSP CASTER SUGAR

PINCH OF SALT

2 TBSP COLD WATER

You can buy ready-toasted sesame seeds, but if you buy them untoasted, put them in a dry frying pan and heat gently, stirring occasionally, until lightly browned.

To make the pastry, preheat the oven to 240°C/475°F/ Gas 9. Put the flour in a bowl, add the butter and rub in using your fingertips until the mixture resembles breadcrumbs. Mix in the sesame seeds, sugar and salt. Sprinkle the water over the top and mix roughly into the dry ingredients with the fingertips of one hand.

Tip the mixture into a ceramic pie dish (about 23cm/9in in diameter) and press down to form an even layer over the bottom of the dish. Prick all over with a fork and bake for 5 minutes. Take out of the oven and leave to cool slightly.

Meanwhile, put all the ingredients for the filling — except the sesame seeds — into a mixing bowl and stir well to combine.

Turn the oven down to 180°C/350°F/Gas 4. Spoon the filling over the pastry, smoothing with the back of the spoon. Sprinkle the sesame seeds over the top.

Put the pie back in the oven and bake for 40 minutes, turning it around halfway through cooking to ensure it cooks evenly.

Take the pie out of the oven and leave to cool for 10 minutes. Serve warm or at room temperature. You can store this in the fridge, but take it out 30 minutes before serving to remove the chill.

BLACKCURRANT AND HONEY JELLY

The taste of honey spreads beautifully through this fruity number. A jelly is a great vehicle for flavour, but remember that a lower temperature subdues flavour and you will be eating this cold, so be prepared to add more honey or crème de cassis.

It is best to use a large jelly mould: a wobbly shining jelly looks so glamorous on the table. But the recipe also works well in six glasses (200–250ml/7–9floz) and this does avoid the stress of unmoulding a jelly in the middle of entertaining and also means you could use slightly less gelatine for an even more tender texture (I've given the minimum amount for the jelly to stand safely on a plate). Make this the day before you serve it.

SERVES 6

10 SMALL SHEETS (20G/¾OZ) OF LEAF GELATINE

750ML/26FL OZ PRESSED APPLE JUICE

5 TBSP HONEY (LIGHT OR MEDIUM)

4 TBSP CRÈME DE CASSIS

5 TBSP BLACKCURRANTS, OR A MIX OF BERRIES (THAWED FROZEN ONES ARE FINE)

Put 3 sheets of gelatine in a shallow dish, cover with cold water and leave to soak for 5 minutes.

Meanwhile, pour 250ml/9fl oz of the apple juice into a pan and add 2 tbsp honey. Heat gently, stirring occasionally, until the liquid is warm and the honey has dissolved. Squeeze the water out of the gelatine, add to the apple juice and stir thoroughly for 30 seconds, or until the gelatine dissolves completely. Do not get the gelatine too hot or the jelly won't set. Pour the liquid into a jelly mould (about 1.2 litres/2 pints capacity) or six glasses. Leave to cool, then chill.

Once the jelly has set, put the remaining 7 sheets of gelatine in a shallow dish of cold water and leave to soak for 5 minutes. Pour the rest of the apple juice into a pan. Add the crème de cassis and the remaining 3 tbsp honey. Stir well as you heat the mixture gently. Taste and add more crème de cassis and honey if you like. Squeeze the water out of the gelatine and stir it into the warm liquid until it dissolves completely. Stir in the fruit. Pour into the mould on top of the set jelly. The fruit will rise to the top to form a layer. Cool quickly by putting the mould in a bath of ice-cold water. Once cold, chill for a couple of hours, or overnight, until the jelly has set.

To turn out, gently run the tip of a knife around the top of the jelly to loosen it, then dip the mould into a bowl of very hot water for 5 seconds. Put a serving plate on top; if you wet the plate first, you can easily slide the jelly into position. Invert the jelly onto the plate and tap the plate and jelly down firmly on a work surface. Should the jelly not turn out well, just spoon it into bowls and top with more fruit and perhaps some single cream.

HONEY-BUTTERED FRENCH PEACHES

This is my version of a recipe in The Art of Cooking with Vegetables *by Alain Passard, the French chef who champions fruit and vegetables at L'Arpège, his three-star restaurant near the Musée Rodin in the 7th arrondissement in Paris. Passard literally puts plants at the centre of the table – using them as table decorations as well as at the heart of his dishes. Fresh produce is picked at his three farms and transported to Paris on the same day, shortly before cooking, for maximum freshness and taste.*

The recipe sings a honeyed song of the South, partnering honey with peaches. Use ripe fruit if you can, but slightly firmish ones work OK too, which is useful.

Passard uses grenadine, which makes this beautiful dish even sunnier. Add a tablespoon to the initial mix if you have it in your cupboard. His stroke of genius is to add olive oil at the end; he uses much more olive oil – and butter – than I do. I also like to add basil, but this is optional.

SERVES 4–6

4 PEACHES

1 UNWAXED LEMON

20G/¾OZ SALTED BUTTER

GOOD PINCH OF SAFFRON STRANDS, SOAKED IN A LITTLE WARM WATER

2½ TBSP LIGHT HONEY (PASSARD RECOMMENDS ACACIA)

1 TBSP GRENADINE (OPTIONAL)

5 BASIL LEAVES (OPTIONAL)

1½–2 TBSP EXTRA VIRGIN OLIVE OIL

Cut the peaches in half, then cut each half into three segments, leaving the skin on and pulling away the stone. Cut the lemon into six wedges, top to bottom, and remove any visible pips.

Melt the butter in a large shallow pan and stir in the saffron and its soaking liquid, the honey and grenadine, if using. Place the fruit in the pan, ideally in a single layer. Cook over a low heat for 20 minutes or so, until the fruit has softened but still has its shape. You may need to move the fruit around gently from time to time.

If you want to add basil, chop the leaves fairly fine, scatter over the peaches and lemons and give them a quick stir.

Serve warm or at room temperature, drizzling the oil over the peaches just before serving. This is best made an hour or so before sitting down to the main course. If the butter has solidified slightly, gently heat the dish before serving.

BUTTERMILK PANNACOTTA WITH HONEYCOMB

Honeycomb is one of nature's treats: a gleaming segment cut from the frame in a hive, honey at its most natural. Bees store their honey here in hexagonal wax containers, as if in thousands of little pots.

It is quite difficult to buy large pieces of comb. You are most likely to find them in delis, farm shops, honey specialists and gift shops with links to a local beekeeper. Seek them out: honeycomb is so beautiful that you can make a pudding or breakfast special with just a small piece. I also love it on crumpets for tea.

Buttermilk was traditionally the liquid left over from butter-making. Now it is made specifically for its slightly tangy flavour, which works well in a creamy pannacotta alongside the honeycomb. The comb is the honey component of the dish, and the pannacotta is sweetened with sugar. Honey is a precious and powerful spoonful, best used carefully and without overloading its flavour.

You can make this the day before you serve it.

SERVES 6

2 SMALL SHEETS (3.2G) OF LEAF GELATINE

300ML/10FL OZ DOUBLE CREAM

250ML/9FL OZ BUTTERMILK

100G/3½OZ CASTER SUGAR

1 VANILLA POD, SPLIT IN HALF LENGTHWAYS AND SEEDS SCRAPED OUT,
OR 1 TSP VANILLA EXTRACT

6 THIN SLICES OF HONEYCOMB

1 RIPE MANGO, PEELED AND CUT INTO SMALL SLICES, OR 150G/5½OZ BERRIES,
SUCH AS RASPBERRIES OR BLUEBERRIES (OPTIONAL)

Put the sheets of gelatine in a shallow dish, cover with cold water and leave to soak for 5 minutes.

Meanwhile, pour the cream and buttermilk into a pan. Add the sugar and either the seeds of the vanilla pod or the vanilla extract. Heat gently, stirring often, until the sugar dissolves.

Squeeze the water out of the gelatine and stir it into the warm cream mixture until it dissolves completely. Pour the liquid into six 100ml/3½fl oz moulds (small coffee cups will do). Leave to cool, then cover with clingfilm and chill for at least 1 hour. The pannacotta will set completely as it chills.

To serve, run the tip of a knife around the top of each mould, then briefly dip the mould into a bowl of hot water and turn out onto a plate. Serve with a slice of honeycomb on the side, along with mango or berries if you like.

FIG AND HONEY FRANGIPANE TART

Honey is used both to flavour the almond base and as a glaze. Fresh or dried figs work equally well in this tart and sometimes I use both together, especially when fresh figs are expensive.

My favourite honey for this dish is a fragrant one such as heather honey. Use plenty of honey to glaze, so the colours of the tart gleam.

SERVES 6

8 FRESH OR DRIED FIGS (OR 4 FRESH AND 4 DRIED)

60G/2¼OZ UNSALTED BUTTER, AT ROOM TEMPERATURE

3 TBSP HONEY (A FRAGRANT HEATHER OR GREEK HONEY WORKS WELL), PLUS 2½ TBSP HONEY, TO GLAZE

2 TABLESPOONS BRANDY OR RUM

2 EGGS

50G/1¾OZ GROUND ALMONDS

2 TBSP PLAIN FLOUR

SINGLE CREAM OR CUSTARD, TO SERVE

PASTRY

70G/2½OZ PLAIN FLOUR, PLUS EXTRA FOR DUSTING

PINCH OF FINE SEA SALT

30G/1OZ COLD UNSALTED BUTTER, CUT INTO SMALL CUBES

1 EGG, BEATEN

First make the pastry. You can do this quickly in a food processor by pulsing the flour, salt and butter until the mixture resembles breadcrumbs, then adding the egg and pulsing until the dough comes together in a ball, adding a little cold water if necessary. To make the pastry by hand, put the flour and salt in a mixing bowl and rub in the butter until the mixture resembles breadcrumbs. Mix in the egg using a round-bladed knife, adding a little cold water if needed, to bring the dough together into a ball. Wrap the dough in clingfilm and chill for 30 minutes.

If using dried figs, soak them in warm water for 30 minutes.

Heavily dust a work surface with flour. Dust your rolling pin and the top of the dough and roll out to form a large thin disc, about 25cm/10in diameter, giving it a quarter-turn every few rolls to keep it even. Use the rolling pin to carefully lift the pastry into a 20cm/8in round loose-bottomed tart tin.

Gently push the pastry into the sides of the tin and trim off the edges. Chill for 20 minutes.

Meanwhile, preheat the oven to 180°C/350°F/Gas 4. Put a baking sheet into the centre of the oven. You do not precook this pastry and the heat of the baking sheet will help the bottom to cook through.

Mix the butter and honey together using a hand-held electric whisk or a food processor – or by beating thoroughly with a wooden spoon. Add the brandy and eggs and stir in well. Fold in the ground almonds and flour. Dollop spoonfuls of the mixture into the pastry case and spread evenly.

Cut the figs in half, removing any stalks, and arrange on top of the frangipane mixture.

Bake for 25 minutes. Turn down the oven to 160°C/325°F/Gas 3 and cook for a further 15–20 minutes. If the pastry looks as if it is burning, cover the edge of the tart with foil. Leave to cool in the tin for 10 minutes.

Meanwhile, gently heat the honey for the glaze. When the tart is slightly cool, remove from the tin and put on a serving plate. Brush the honey generously all over the top to give it a gleaming finish.

Serve warm or cold, with single cream or crème anglaise, flavoured with a touch of brandy, if you like.

PEARS POACHED IN MEAD

Mead is made from fermented honey; it was mankind's earliest form of booze and very much part of our love affair with honey. When made well, and not cloyingly sweet, mead is a truly lovely drink that is superb alongside puddings, blue cheese and rich pâtés, in much the same way as Sauternes.

The simplicity of the ingredients in this recipe allows all the elements to shine. I've used just part of a bottle of mead so there's the rest to drink. This dish is best made the day before serving, to allow the pears to absorb the flavours and gloss of the syrup.

Poaching is a useful way of softening less-than-ripe pears. You can't always wait for this obstinate fruit to ripen to perfection in the fruit bowl. In any case, totally ripe pears are wonderful to eat simply cut into quarters and served with honey-drizzled cheese, be it soft, hard, sheep's, cow's, goat's, plain or blue.

SERVES 6

375ML/13FL OZ MEAD

30G/1OZ HONEY (A LIGHT HONEY SUCH AS WILDFLOWER OR ORANGE BLOSSOM)

30G/1OZ CASTER SUGAR

6 PEARS, PEELED, STALK LEFT ON

2 BAY LEAVES

SINGLE CREAM, TO SERVE

Put the mead in a large, shallow, lidded sauté pan with the honey and sugar. Heat gently, stirring occasionally, until the sugar and honey have dissolved.

Place the pears in the pan on their sides. Tuck the bay leaves in among the fruit. Cover with a lid and turn up the heat until the liquid comes just to the boil. Turn down the heat so the liquid simmers, cover the pan and cook the pears for 10 minutes, then carefully turn them over using two spoons and cook for another 10 minutes. Continue to simmer, turning every 10 minutes, until the pears are tender all the way through (hard pears take about 50–60 minutes).

Take the pears out of the pan and boil the liquid for 5 minutes or so, reducing it until slightly syrupy. You can serve the pears straight away or, better still, leave the pears in the syrup in a covered serving bowl for a few hours or overnight in the fridge, turning them over once or twice more.

Serve the pears with the syrup and single cream and perhaps a biscuit, such as pollen shortbread (page 125).

BAKLAVA

Many masters of baklava sell their pastries on the Uxbridge Road in West London, near where I live. Having acquired a serious baklava habit, I became curious about this great honey classic and started to make my own. My biggest discovery was that baklava are best made a few days before eating, as the honey-syrup soaks into the nuts and pastry over time.

This recipe is a good one for the baklava novice because it is made in a small square tin, which makes the filo easier to handle. I add a fair amount of lemon juice to make the flavour slightly fresher and less super-sweet than some commercial products.

MAKES 16 SMALL SQUARES
100G/3½OZ UNSALTED BUTTER

ABOUT 280G/10OZ FILO PASTRY (6 LARGE SHEETS)

150G/5½OZ NUTS, SUCH AS PISTACHIOS, WALNUTS, HAZELNUTS OR ALMONDS (JUST ONE KIND, OR HALF AND HALF), FINELY CHOPPED

HONEY SYRUP
150ML/5FL OZ WATER

100G/3½OZ CASTER SUGAR

2 TBSP HONEY (A LIGHT FRAGRANT ONE SUCH AS ORANGE BLOSSOM IS BEST)

JUICE OF ½ LEMON

½ TBSP ORANGE FLOWER WATER OR ROSEWATER

First make the syrup: put the water and sugar in a small pan, bring to the boil and boil for 5 minutes. Turn off the heat and stir in the honey, lemon juice and orange flower water. Give the mixture a quick stir from time to time so that the honey dissolves fully. Pour the syrup into a jug and leave to cool.

Preheat the oven to 160°C/325°F/Gas 3 and put a baking sheet on the centre shelf. Melt the butter in a small pan over a medium–low heat.

Unroll the filo pastry so the six sheets lie on top of each other on the work surface. Cut them in half and then trim each pile (if necessary) to get a square roughly the size of a shallow 20cm/8in square baking tin.

Brush the bottom and sides of the tin generously with some of the melted butter. Lay one square of filo pastry in the tin

and brush the top well with butter. Repeat with the remaining five pieces from that pile, brushing each piece with butter before laying the next one on top.

Spread the chopped nuts over the filo pastry (if you like, hold back 2 tbsp of nuts to scatter on top after cooking). Layer the next six squares of filo pastry on top of the nuts, starting with a piece that is buttered on one side and placed butter-side down on the nuts; then butter this sheet on the top and continue layering up the pastry. Finish with a final brushing of butter. It seems a lot of butter, but you need to use it all in order to get the right texture and cohesion.

Use a sharp knife to cut all the way through the layers to make 16 small squares. Put the baklava in the oven on the hot baking sheet and bake for about 45 minutes, or until the top is brown.

Pour the syrup slowly and carefully over the hot baklava, making sure it goes into the cracks between the pieces. If you've kept some back, scatter over the remaining nuts. Leave to cool, so the honey-syrup thickens. Ideally leave for a day or two, and best of all 4 days, before eating.

SICILIAN HONEY BALLS (SFINGI)

The Sicilians have many sublime honeys and a great tradition of using them in sweet recipes. Sicilian food is especially fascinating because it contains elements of all the cultures that have occupied this island over the centuries: ancient Greek and Roman, Arab, Norman, Spanish, French and now Italian. These little puffy balls, called sfingi, *could be said to have Arabic roots, although the choux pastry feels Bourbon, too.*

Sfingi are traditionally made to celebrate the feast day of San Giuseppe, or St Joseph, on 19 March. While attending such a feast, in the south-west of the island, I found one of the most delicious honeys I've ever eaten. It was sold, without a label, from a tabletop beside an organic lemon grove and has the richest gold of any of my honey pots, as well as a great flavour. Sometimes I put it on my kitchen table just to brighten the room.

SERVES 6

85G/3OZ UNSALTED BUTTER

PINCH OF SALT

1 TSP CASTER SUGAR

225ML/8FL OZ WATER

100G/3½OZ PLAIN FLOUR

½ TSP FINELY GRATED ORANGE OR LEMON ZEST
(USE UNWAXED FRUIT)

3 EGGS

VEGETABLE OIL FOR DEEP-FRYING

HONEY FOR DRIZZLING (IDEALLY A LIGHT AND FRAGRANT ONE, SUCH AS SICILIAN ORANGE BLOSSOM OR ACACIA)

Put the butter, salt, sugar and water in a heavy-bottomed pan. Bring to the boil over a medium–high heat so the butter melts and the sugar dissolves.

Add the flour and citrus zest and stir hard, using a wooden spoon, until the mixture forms a ball. Take off the heat and leave to cool for a minute. Use a hand-held electric whisk to beat in the eggs, one at a time. Beat for a few minutes until the mixture is smooth and shiny.

Heat 5cm/2in of oil in a wide, deep pan until it reaches 180°C/350°F, or until a cube of bread browns in 30 seconds. Use two teaspoons to scoop teaspoonfuls of the mixture into the hot oil, frying about 8–10 pieces at a time and turning them over as they brown. Lift the cooked balls out of the oil with a slotted spoon and drain on kitchen paper.

These are best eaten as soon as they are cool enough to put in your mouth, drizzled generously with honey.

HONEY CHEESECAKE

This sophisticated dessert is adapted from one in Cooking at the Merchant House *by Shaun Hill, who is a scholar of classical Greece and its food as well as a great chef. This is a modern dessert but I got the idea from the honey cheesecakes that were much eaten by the ancient Greeks, as lyrically referred to in Euripides in the fifth century* BC: *'Cheese-cakes, steeped more thoroughly /In the rich honey of the golden bee.'*

SERVES 8

400G/14OZ CREAM CHEESE

100ML/3½FL OZ SOURED CREAM

100G/3½OZ SOFT BROWN SUGAR

3 EGGS

50G/1¾OZ HONEY (IDEALLY A FRAGRANT MEDIUM OR DARK HONEY SUCH AS THYME, HEATHER OR BUCKWHEAT)

1 TSP VANILLA EXTRACT

FINELY GRATED ZEST OF 1 UNWAXED LEMON

HAZELNUT BASE

115G/4OZ DIGESTIVE BISCUITS

30G/1OZ TOASTED HAZELNUTS, CHOPPED

60G/2¼OZ UNSALTED BUTTER, PLUS EXTRA FOR GREASING

1 TBSP CASTER SUGAR

TOPPING

300ML/10FL OZ SOURED CREAM

1 TSP VANILLA EXTRACT

3 TBSP RUNNY HONEY (IDEALLY TWO OR THREE KINDS: LIGHT, MEDIUM AND/OR DARK)

First make the hazelnut base. Preheat the oven to 200°C/400°F/Gas 6. Butter a 20cm/8in springform cake tin. Put the biscuits and nuts in a food processor and whizz until you have reasonably fine crumbs. Alternatively, put the biscuits in a plastic bag and bash with a rolling pin to make fine crumbs. Tip the mixture into a mixing bowl. Melt the butter and stir into the crumbs, along with the sugar. Press the mixture into the tin, aiming to get an even layer with a lip at the edge. Bake for 7 minutes, then leave to cool.

To make the filling, preheat the oven to 180°C/350°F/Gas 4. Put the cream cheese and soured cream in a mixing bowl and stir until well combined. Stir in the sugar and then the eggs, one at a time. Stir in the honey, vanilla and lemon zest.

Pour the mixture over the cooled base in the cake tin. Bake for about 30 minutes, or until the mixture is firm in the middle. Take out of the oven and leave to cool for 10 minutes.

Turn up the oven to 200°C/400°F/Gas 6. To make the topping, mix together the soured cream and vanilla and spread over the cheesecake. Put in the oven for 5 minutes, or until a glossy top has formed. Leave to cool for at least 2 hours. You can make this the day before eating and leave it in the fridge, covered, overnight.

Run the tip of a hot knife around the top of the cheesecake and carefully open the springform tin. If the cheesecake splits, just fill in the cracks with a little soured cream.

Put the cheesecake on a serving plate. If you want to remove the bottom of the springform tin, slide a large, hot knife between the cheesecake and the metal, then use a large fish slice to transfer the cheesecake to the serving plate.

Decorate the top of the cheesecake just before serving. You can just use one kind of honey, but two or three kinds looks better. Drizzle the light and then the darker honey over the top of the cheesecake and marble them together by running the back of a spoon in a wavy pattern over the honeys.

Serve the cheesecake on its own or with fruit such as fresh raspberries.

HONEY AND CARDAMOM KULFI

Traditionally, kulfi is made by simmering milk gently for several hours until much of the liquid has evaporated. The shortcut is to use condensed and evaporated milk. Either way, the sweet intensity of the mixture means the ice stays smooth even though it hasn't been churned in an ice-cream maker. You can adapt this recipe by adding chopped pistachios or flavourings such as rose water.

The Indian nation has a collective sweet tooth, with many desserts and sweets made mostly with sugar and sometimes with honey. Anil Kishore Sinha in his wonderful Anthropology of Sweetmeats *quotes a passage from the ancient Rigveda verses evoking an early love of honey: 'They would drive their chariot around and in a leisurely manner while drinking honey, listening to the beautiful hummings of bees and riding on the chariot also humming like bees...'*

Indian desserts are often made for special occasions, and I like to bling-up the kulfi using the glittery baubles sold for baking. But you can also eat them plain. They are useful to have in the freezer for those moments when you suddenly want a special sweet dish.

SERVES 6

3 CARDAMOM PODS

150ML/5FL OZ EVAPORATED MILK

150ML/5FL OZ CONDENSED MILK

300ML/10FL OZ DOUBLE CREAM

3 TBSP HONEY (THIS CAN TAKE A STRONG HONEY SUCH AS CHESTNUT OR HEATHER)

SUGARY BLING, SUCH AS SILVER BALLS OR GOLD LEAF (OPTIONAL)

Use a small sharp knife to slit open the cardamom pods. Remove the seeds and crush with a mortar and pestle. Put the milks, cream and honey in a mixing bowl, add the cardamom seeds and stir well to combine.

Pour the mixture into six small moulds. You can buy special conical kulfi moulds with screw-on lids, but I use metal dariole moulds with a capacity of around 100ml/3½fl oz. Cover the moulds with clingfilm and put into the freezer for at least 3 hours.

When you are ready to eat, slide a small, sharp knife around the top of each kulfi. Dip the bottom of the mould in a bowl of just-boiled water for 10 seconds or so, then turn the kulfi out onto a plate. If you like, decorate each ice with silver balls or even, if you are feeling flash, gold leaf.

BROWN BREAD ICE CREAM

Brown bread and honey are delicious as honey sandwiches. They also go together in this unusual ice cream. Brown bread ice cream sounds like a quirky modern combination, but the recipe goes back to at least the 19th century, when it was said to be one of Queen Victoria's favourite desserts. The flavour and crunchy texture of toasted brown breadcrumbs make this feel like a substantial dessert without losing the essential delight of an ice.

This particular recipe comes from Charlotte Dunne, a trained cook who teaches beekeeping to children at Ashley Church of England Primary School in Walton-on-Thames in Surrey. I went to inspect the hives with Charlotte and a group of 6- and 7-year-olds dressed in beesuits. The kids were impressively confident about approaching the bees. They knew all about pollination and the importance of these insects to the natural world, from being with the bees and looking after the fruit trees and vegetables that are grown in the school grounds. These children were absorbing big lessons in the best possible way.

SERVES 10

125G/4½OZ BROWN BREADCRUMBS

125G/4½OZ DEMERARA SUGAR

600ML/20FL OZ WHIPPING OR DOUBLE CREAM

2 TBSP ICING SUGAR

3 TBSP WILDFLOWER RUNNY HONEY

1 TBSP DARK RUM, OR A LIQUEUR SUCH AS COINTREAU

Preheat the oven to 190°C/375°F/Gas 5. Spread the breadcrumbs over a baking sheet. Scatter over the sugar and mix them together with your fingertips. Place the baking sheet in the oven for about 10–12 minutes, until the crumbs are starting to brown and the sugar to melt, keeping a close eye that they do not burn. Turn over the crumbs and continue to watch closely, turning the crumbs at regular intervals until they are brown all over and the sugar is caramelizing. Remove any crumbs that have burnt, as these will taste bitter. Leave to cool, then break up any large clumps of sugar and crumb.

In a large bowl, lightly whip the cream and fold in the icing sugar, followed by the cooled crumbs, and then fold in the honey and rum or liqueur. Don't be tempted to add extra booze because alcohol will prevent the mixture from freezing.

Freeze the mixture in an ice-cream maker. Alternatively, put the mixture into a shallow lidded container, about 23 x 18cm/9 x 7in. Freeze for about 2 hours, until ice crystals form around the edges. Remove from the freezer and beat well with a whisk or fork until slushy. Freeze for a further 2 hours and repeat the process.

Remove the ice cream from the freezer 20 minutes before serving. Serve on its own or with fruit.

CHOCOLATE AND CHESTNUT HONEY ICE CREAM

Chocolate and honey are a wonderful flavour combination and ice cream is a great medium for this flavour: the melting cream unrolls the honeyed warmth and the rich chocolate onto your tongue. No wonder ice cream is so lickable.

Chilling subdues flavour and it is best to use a strongly flavoured honey for this ice. Not everyone likes chestnut honey; some positively dislike its bitter notes. I'm a fan and love the way the honey's almost resinous aromatic quality comes through when paired with good dark chocolate. You can happily use another honey here, but I'd recommend another strongly aromatic one, such as heather or Greek mountainside honey.

SERVES 8–10

150G/5½OZ DARK CHOCOLATE (AT LEAST 70% COCOA SOLIDS), BROKEN INTO PIECES

300ML/10FL OZ FULL-FAT MILK

300ML/10FL OZ DOUBLE OR WHIPPING CREAM

3 EGG YOLKS

1 TBSP DARK MUSCOVADO SUGAR

115G/4OZ CHESTNUT HONEY, PLUS OPTIONAL EXTRA FOR DRIZZLING

50G/1¾OZ PECAN NUTS OR HAZELNUTS

Put the chocolate, milk and cream in a pan and heat very gently, stirring occasionally, until the chocolate has melted and the mixture is nearly boiling. Turn off the heat and leave to cool slightly.

In a bowl, whisk together the egg yolks, sugar and honey. Pour the cooled chocolate cream into the bowl, whisking all the time. Pour the mixture back into the pan. Cook over a very low heat, ideally on a heat diffuser, stirring almost constantly, for about 10 minutes, or until the mixture thickens slightly. Cool, then chill in the fridge. Freeze the chilled mixture in an ice-cream maker. Alternatively put the mixture into a shallow lidded container and freeze for about 2 hours, until ice crystals form around the edges. Remove from the freezer and beat with a fork, mixing the crystals thoroughly back into the mixture. Repeat three times, then leave to freeze completely.

To toast the nuts, preheat the oven to 150°C/300°F/Gas 2. Put the nuts on a baking sheet in one layer and cook for 15 minutes; after 10 minutes, check they aren't burning and turn the baking sheet round to ensure they toast evenly. Leave to cool, then chop roughly.

About 30 minutes before serving, transfer the ice cream from the freezer to the fridge. To serve, scoop the ice cream into small dishes – it is an intense ice, so 2 small scoops will suffice. Scatter with the nuts. You can drizzle extra chestnut honey on top, but I find there's enough honey flavour in the ice cream itself.

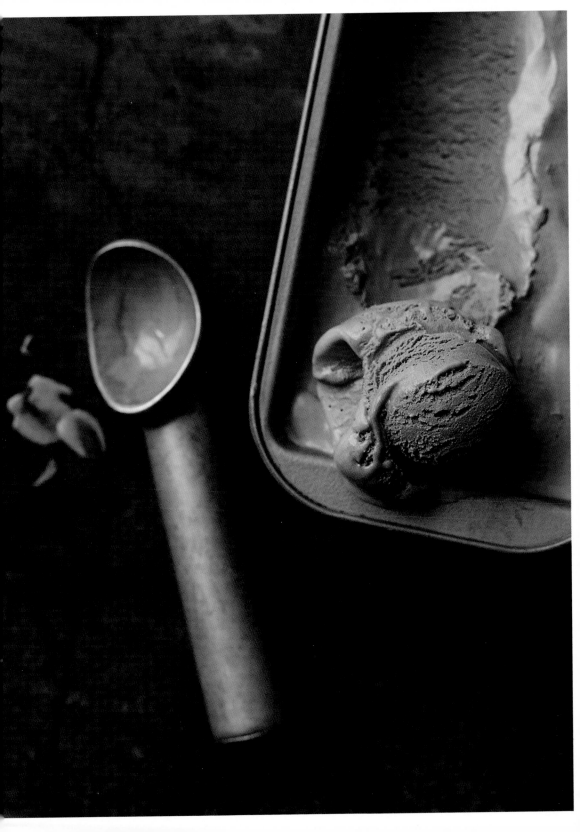

PRESERVES, SWEETS AND DRINKS

HONEY, APPLE AND ROSEMARY JELLY

This jelly spans the sweet and the savoury. You can eat it on toast or scones or put a spoonful on your plate to accompany roast pork or cheese on toast; it is great with cold meats and all cheeses. It also works well as a glaze on an apple or other fruit tart. Gently warm the jelly in a small pan and brush it liberally over the finished tart, then leave to set. The lemon juice heightens the flavour and helps the pectin, a natural setting agent in the apples, to work more effectively. The honey means this jelly isn't as crystal clear as some, but the preserve develops a beautiful deep pinky-orange colour as the mixture boils, and it looks divine.

MAKES 3 X 350ML/12FL OZ JARS

1KG/2LB 4OZ COOKING APPLES (ABOUT 4 LARGE APPLES)

1.2 LITRES/2 PINTS WATER

2 LARGE SPRIGS OF ROSEMARY

ABOUT 500G/1LB 2OZ GRANULATED SUGAR

ABOUT 300G/10½OZ HONEY (A LIGHT, FLORAL TYPE
SUCH AS CLOVER OR WILDFLOWER IS BEST)

2 LEMONS

Cut the apples into quarters. Cut out the stalks and any blemishes. Roughly chop the apples and put into a large pan — skin, core, pips and all. Pour in the water and add the rosemary. Bring to the boil, then turn down the heat and simmer for 45 minutes. Ladle the mixture into a jelly bag suspended over a large bowl. Leave this to drip through the bag without squeezing it (3–4 hours is enough). Measure out the liquid and transfer to a pan. For every 500ml/18fl oz of liquid, add 250g/9oz sugar, 150g/5½oz honey and the juice of 1 lemon. Bring the mixture to the boil, stirring occasionally to ensure the sugar has dissolved by the time it is boiling. Meanwhile, put a saucer in the freezer.

Boil the jelly mixture for about 10 minutes, then put a teaspoon of the liquid on the cold saucer and put in the fridge for 1 minute. Push the mixture with your finger and if it wrinkles, then it has reached setting point. This is known as the 'wrinkle test'. If not, then continue to boil the mixture and test it every few minutes, cleaning the saucer and putting it into the fridge between times. It can take 15 minutes or so, depending on the balance between pectin, acid and sugar. Turn off the heat and leave to settle for 10 minutes.

While the jelly mixture is boiling, sterilize the jars. Thoroughly wash 3 x 350ml/12fl oz pots or 2 x 500ml/18fl oz pots (you need about 1 litre/1¾ pints capacity in total) and their lids. Put them in the oven, set it to 110°C/225°F/ Gas ¼ and leave for 10 minutes. Skim off any scum from the surface of the jelly; ladle the jelly into the hot sterilized jars, filling them right to the top. Put the sterilized lids on immediately and leave to cool.

The jelly will set in the jar as it cools. Keep in a cool, dark place and eat within 1 year. Once opened, keep in the fridge.

HONEY AND WHISKY TRUFFLES

These dense, sumptuous truffles are straightforward to make and provide a special after-dinner treat. Try this recipe with different chocolates and honeys and you will find significant variations in flavour. Heather honey is a good choice, not least to accompany its compatriot, Scotch whisky. I've also used chestnut honey with brandy and Greek thyme honey with rum to good effect. The Scottish liqueur Drambuie will emphasize the honey flavour.

MAKES ABOUT 30

200G/7OZ DARK CHOCOLATE (AT LEAST 70% COCOA SOLIDS), BROKEN INTO PIECES

100G/3½OZ UNSALTED BUTTER

3 TBSP DOUBLE CREAM

2 TBSP WHISKY OR OTHER SPIRIT

2–3 TBSP HONEY (A DARK, FRAGRANT HONEY SUCH AS HEATHER IS BEST)

UNSWEETENED COCOA POWDER, TO COAT

Put the chocolate in a heatproof bowl with the butter, cream and whisky. Place over a pan of simmering water (the bottom of the bowl shouldn't touch the water) and heat gently, stirring occasionally, until the chocolate has melted and everything has come together. It is important that you do not overheat the bowl, so make sure that the water doesn't bubble up. If the chocolate gets too hot, it will 'seize' and separate from the fat. (If the mixture does separate, cool the bowl by putting the bottom in cold water and then place in the fridge for about 10 minutes. As the mixture cools, you can stir the fat into the chocolate with a fork. The texture of the truffles won't be as good, but the taste will be fine.)

Stir the honey into the melted mixture, to taste, remembering that flavours get less strong as they cool and chill. The amount of honey will depend on what kind of honey, chocolate and, to a lesser extent, spirit you are using and how they balance together – any excuse to keep tasting.

Leave the mixture to get cold, then cover with clingfilm and put in the fridge for at least 2 hours to harden.

Cover the bottom of a small plate with cocoa. Take a teaspoonful of the truffle mixture and use your hands to mould it into a rough truffle shape – these are truffles as if from the earth, not the perfect circles of the confectioner. Roll the truffle around in the cocoa, then transfer to a plate. Repeat with the rest of the mixture.

These will keep in the fridge, covered, for a week or two.

HONEY FUDGE

A wickedly sweet indulgence; just a small square or two makes a good treat with a cup of coffee after lunch, giving a burst of sugar-energy to get you up from the table and into the afternoon.

Honey is a beautiful flavouring for fudge. I'd recommend a medium or dark honey with a bit of power in its flavour, but even a light honey will come through well.

MAKES 25 SQUARES

400G/14OZ GRANULATED SUGAR

400G/14OZ CONDENSED MILK

100ML/3½FL OZ WATER

PINCH OF SALT

3 TBSP HONEY (TRY A FLAVOURFUL DARK OR MEDIUM HONEY, SUCH AS HEATHER OR FOREST HONEY)

1 TSP VANILLA EXTRACT

50G/1¾OZ COLD UNSALTED BUTTER

Lightly oil a 20cm/8in square shallow baking tin or line it with greaseproof paper.

Put the sugar, condensed milk and water in a large, heavy-bottomed saucepan. The mixture will froth up, so be sure your pan is large enough. Heat the mixture gently, carefully drawing a wooden spoon through the sugar until it has completely dissolved; this takes about 5 minutes or more, if done gently. Stir in the salt and 2 tbsp of the honey.

Put a sugar thermometer in the pan and bring to the boil. Boil vigorously, stirring almost continuously to ensure the mixture doesn't burn on the bottom of the pan. Some of it will go very dark brown, but don't worry – just keep stirring. Boil until the mixture darkens and thickens and reaches the soft ball stage (about 116°C/240°F on the thermometer, or when a little of the mixture dropped into cold water forms a soft ball); this will take around 15–20 minutes.

Turn off the heat and stir in the vanilla extract, butter and the remaining 1 tbsp honey. Beat hard with a wooden spoon. As the fudge cools, it will become thick and granular. You can speed this up by putting the pan in a shallow tray of cold water. Working quickly, as the fudge is hardening all the time, spoon the mixture into a tin, smoothing the top with the back of a spoon (if the mixture is hardening fast, heat the spoon by dipping it briefly in a bowl of just-boiled water).

Leave the fudge to cool slightly, then mark into 25 squares and leave to set completely. This keeps well in an airtight container, but is best eaten within a few weeks.

CHEWY HONEY BITES

Bees Abroad (www.beesabroad.org.uk) is an excellent charity that helps beekeepers in developing countries. Volunteers help beekeepers to set up small businesses using indigenous bees and techniques appropriate to their setting. Countries including Uganda, Kenya, Cameroon, Malawi and Ghana have benefited from the charity's work. The charity sets up its stall at agricultural and honey shows to raise money and awareness.

These honey bites are a slight adaptation of a recipe from Bees Abroad. They are quite hard to begin with, and similar to biscotti. Their texture softens over time and the honey flavour develops. They are great with a cup of coffee after a meal or to take on a walk for a quick fix of honey energy.

MAKES 25 SQUARES
100G/3½OZ UNSALTED BUTTER, PLUS EXTRA FOR GREASING

225G/8OZ HONEY

70G/2½OZ SOFT BROWN SUGAR

225G/8OZ PLAIN FLOUR

1 TSP BAKING POWDER

60G/2¼OZ CHOPPED NUTS, SUCH AS MACADAMIA

100G/3½OZ DRIED FRUIT

SPICE MIX
½ TSP GROUND GINGER

1 TSP GROUND CINNAMON

⅛ TSP GROUND CLOVES

SEEDS FROM 6 CARDAMOM PODS, CRUSHED

Preheat the oven to 180°C/350°F/Gas 4. Butter a 20cm/8in square tin.

Put the butter, honey and sugar in a small pan and warm gently until melted.

Put the flour and baking powder in a mixing bowl. Stir in the spices, then mix in the chopped nuts and dried fruit. Pour the melted butter mixture into the bowl and stir well to combine. Spoon the mixture into the prepared tin and bake for 25 minutes.

Leave to cool in the tin, then cut into 25 squares and store in an airtight container for at least 3 days before eating. The bars are hard at first but soften after around a week and will keep for a few months.

NOUGAT

Honey is part of a fragrant family of sweets that came from the Arab world to Europe with the Moors and today can be found in different versions in France, Spain and Italy. There's no denying that making nougat requires careful timing. Please don't embark on this without first reading the instructions all the way through, because you need to be ready to act fast at the crucial moments. You will also need a free-standing electric mixer and a sugar thermometer. Have a go – delicious.

MAKES ABOUT 40–50 PIECES

300G/10½OZ NUTS (EQUAL PARTS HAZELNUTS, ALMONDS AND PISTACHIOS IS A GOOD MIX), OR 300G/10½OZ MIXED NUTS, DRIED FRUIT AND PRESERVED STEM GINGER (FRUIT AND GINGER CUT INTO PIECES THE SIZE OF A RAISIN)

100G/3½OZ HONEY (LIGHT LAVENDER OR ACACIA HONEY ARE TRADITIONAL)

300G/10½OZ CASTER SUGAR, PLUS 15G/½OZ FOR MERINGUE

85G/3OZ LIQUID GLUCOSE

100ML/3½FL OZ WATER

1 LARGE EGG WHITE (ABOUT 40G/1½OZ)

1 TSP VANILLA EXTRACT

Preheat the oven to 160°C/325°F/Gas 3. Put the nuts on a baking sheet and put in the oven for 15 minutes, turning the tin around once or twice if your oven has a slightly uneven heat and checking after 10 minutes that the nuts aren't browning too much. Leave to cool.

Cover a shallow 20–25cm/8–10in square baking tin with baking parchment and, for extra non-stickiness, spray or lightly wipe with a flavourless oil such as sunflower. Put the honey in a pan ready to heat up.

Put the sugar, glucose and water in a large, heavy-bottomed pan. Heat gently, drawing a wooden spoon through the sugar from time to time, until it has completely dissolved. Put a sugar thermometer in the pan, turn up the heat and start to boil vigorously. It will take about 4–6 minutes to reach 140°C/275°F on the thermometer (the soft crack stage, when a little of the mixture dropped into cold water forms threads that bend before breaking). Keep an eye on it, especially once the temperature reaches 120°C/250°F, and in the meantime start to make the meringue.

While the sugar is dissolving, warm the honey in the pan, but do not let it boil. Put the egg white in a free-standing electric mixer and whisk until soft peaks form. Add the remaining 15g/½oz sugar and keep whisking for another minute. Whisk in the warm honey and turn off the machine.

When the sugar syrup has nearly reached 140°C/275°F, start the whisk going again. As soon as the temperature has reached 140°C/275°F, immediately trickle the hot sugar syrup directly onto the egg whites, trying not to get it onto the whisk or the side of the bowl, where it will set. Do this carefully but as quickly as you can.

Keep whisking the egg white mixture hard for about 10 minutes, until thick and glossy. Whisk in the vanilla, then stop the whisk and tip in the nuts (or fruit and nuts). Whisk slowly and briefly to incorporate the ingredients into the ultra-sticky meringue mixture. Stop the machine and give the mixture a final stir with a large metal spoon.

The nougat will get harder as it cools, so work fast. Quickly spoon the nougat into the lined tin. Press the mixture down into the corners and bottom of the tin and smooth the top as best you can with the back of the spoon. Spray or wipe another piece of baking parchment with oil and press it down on the nougat to smooth and compress it. Leave to cool for at least 6 hours or overnight.

Tip the nougat out of the tin onto a chopping board. Carefully peel away the paper. You can then use a serrated knife to cut the nougat into long slices and then into small squares. However, since nougat can be very sticky, I prefer to wrap the whole piece in its parchment and then clingfilm and keep it in the freezer, where it remains firm but not frozen because of its high sugar content. Break off bits shortly before serving.

HONEYED DATES

The rich tradition of medieval Arab cookery includes many sweets made with honey. Recipes from old cookbooks have been explored and translated into modern cooking methods in a remarkable book, Sweet Delights from a Thousand and One Nights: the Story of Traditional Arab Sweets *by Habeeb Salloum, Muna Salloum and Leila Salloum Elias. The dishes in this book are still familiar today, not just in the Middle East but also in countries that came under Arab rule, such as Morocco and Spain. 'Sweets are part of the social culture in the Arab world, where guests are held in high esteem,' explain the authors. Dates are an important staple in Middle Eastern cooking and this dish, adapted from the version in* Sweet Delights*, is a great sweet to pass round with coffee after lunch or supper.*

Stuffing the dates with nuts is not such a long task and the contrast between the dark soft date and the hard white nut is most pleasing. The mixture of saffron, cinnamon, rosewater (use orange blossom water if you prefer) and poppy seeds is exotic without being overpowering, and everything is held together by the sheen and aromatic sweetness of the honey.

MAKES 20

¼ TSP SAFFRON STRANDS

½ TSP ROSEWATER

3 TBSP WATER

20 DATES

20 BLANCHED ALMONDS

1 TBSP HONEY

PINCH OF GROUND CINNAMON

½ TSP POPPY SEEDS

Put the saffron in a small bowl with the rosewater and water. Use a mortar to crush the saffron slightly. Leave to soak while you prepare the dates.

Slit each date and remove the stone. Put an almond into each date and gently press the fruit to enclose the almond inside.

Put the saffron and rosewater mixture in a small saucepan. Add the honey and cinnamon. Bring to the boil and boil gently for 1 minute, until the bubbles are larger and slower-bursting and the mixture has thickened slightly (it will thicken more as it cools). Turn the heat down, add the dates and simmer for 2 minutes. Leave to cool, rolling the dates around in the syrup as it cools.

Sprinkle with the poppy seeds and serve at room temperature.

MANUKA, ROSEMARY AND ORANGE CORDIAL

Manuka honey from New Zealand has a worldwide reputation for its health benefits and is used for anything from stomach complaints to leg ulcers. Peter Molan, a Welsh-born Professor of Biological Sciences at the University of Waikato on the North Island of New Zealand, has worked for decades to find scientific proof of the efficacy of honey in general and manuka in particular.

While there are other honeys that are also potent in terms of health (and please feel free to use any kind in this recipe), manuka certainly smells and tastes like it might do you good, with a menthol tang that is strong even for those who like dark honeys. However, partnered with orange and rosemary and a touch of cloves, it makes a cordial that is delicious, refreshing and soothing on the throat.

MAKES 1 X 750ML BOTTLE

3 UNWAXED ORANGES

5 BUSHY SPRIGS OF ROSEMARY

3 CLOVES

700ML/1¼ PINTS WATER

ABOUT 375G/13OZ GRANULATED SUGAR

ABOUT 2½ TBSP MANUKA HONEY

Use a vegetable peeler to pare the peel from the oranges in thin strips. Put into a pan with the rosemary, cloves and water. Bring to the boil, then turn off the heat, cover and leave to cool and infuse for 24 hours.

To sterilize your bottle, wash it well and put it and its cap on a baking sheet in the oven. Heat the oven to 110°C/225°F/Gas ¼ and leave for 10 minutes.

Strain the infused liquid into a measuring jug. Squeeze the oranges and add the juice to the jug. Note the quantity of liquid and pour it back into the pan. For every 600ml/20fl oz of liquid, add 300g/10½oz granulated sugar. Heat gently until the sugar has dissolved, then bring to the boil and boil for 5 minutes.

Stir in the manuka honey (2 tbsp per 600ml). Pour the cordial into the warm sterilized bottle and put on the cap. Keep in a cool, dark place and, once opened, keep in the fridge. Dilute to taste with still or fizzy water, or with tonic.

LAVENDER HONEY LEMONADE

Lavender honey is one of the best in the world. The famous Narbonne honey, much loved for millennia, is partly based on lavender. People bring back jars or plastic tubs of this light sweet honey from holidays in France and then dream of summer all through the winter.

This recipe plays on the association of summer, honey and fragrant lavender. I've made fresh lemonade using various methods and I like the way this one captures the fragrance of a lemon's essential oils while avoiding some of the bitterness you get when the peel is left to soak for a long time. The mixture is less sweet than some and this keeps the drink extra refreshing.

Lavender flowers vary in their fragrance but all will work fine here. What matters is that they haven't been sprayed (the bees aren't too keen on that either). If you don't have any lavender from a garden, you can get hold of some lavender sugar (see Suppliers, page 188) to keep in the kitchen cupboard.

SERVES 4–6

3 TBSP CASTER SUGAR

3 TBSP HONEY (A LIGHT OR MEDIUM HONEY IS BEST; LAVENDER WOULD BE DELIGHTFUL)

5 HEADS OF UNSPRAYED LAVENDER

6 UNWAXED LEMONS

ICE CUBES, TO SERVE

LEMON SLICES, TO SERVE (OPTIONAL)

Boil a kettleful or large pan of water. Put the sugar and honey in a large bowl. Scrunch up the lavender florets and add to the bowl. Scrape a fork all over the lemons to start to release their oils and put them in the bowl. Pour over 1.2 litres/2 pints of the just-boiled water. Stir briefly and leave for 15 minutes.

Use a slotted spoon to remove the lemons from the water. Scratch them again all over with your fork. Put the lemons back in the bowl. Taste the liquid with a clean spoon and decide if there is enough lavender flavour. If so, remove the flowers. Leave the lemons to release their fragrant aromatic oils for another 15 minutes.

Remove the lemons, cut them in half and squeeze the juice. Pour the juice into the bowl and discard the peel. If you have left the lavender in, remove it now. Strain the liquid into a jug and chill.

Once chilled, taste and adjust the flavour with more honey or lemon juice if necessary. Serve in small glasses – a large wine glass would be ideal – with ice cubes and perhaps an extra slice of lemon.

ICED GREEN TEA WITH LEMONGRASS AND ACACIA HONEY

All herbal teas are much improved by a small spoonful of honey because the sweetness brings out their aromatic flavours. I'm a fan of making my own tisanes out of fresh herbs, as here; they have far more fragrance than herbal tea bags.

The idea for this drink came from a Brazilian chef, Mariana Villas-Bôas. Mariana's brother Jerônimo is an expert on the native stingless bees of Brazil. These insects are not as smilingly benign as they sound — stingless bees bite instead. Fair enough: they need to protect their delectable honey, which is slightly less sweet than honeybee honey and wonderfully fresh and fragrant. It is mostly used for medicinal rather than culinary purposes.

Mariana makes a drink by boiling and cooling lemongrass and water, mixing it with honey, then serving it over ice with a slice of lime. I took this idea and combined it with my quest to make a honeyed iced tea.

Green tea is much abused by being made with boiling water, which brings out unpleasant bitter flavours. Make it with warm water, or infuse it for longer in cold water, as here.

---

5 LEMONGRASS STALKS

1 LITRE/1¾ PINTS WATER

3—4 TBSP ACACIA HONEY, OR ANOTHER LIGHT KIND SUCH AS
ORANGE BLOSSOM OR WILDFLOWER

JUICE OF 1 LIME, PLUS LIME SLICES TO SERVE

2 TSP GOOD-QUALITY LOOSE-LEAF GREEN OR OOLONG TEA

Finely chop the lemongrass and put in a small pan along with the water. Cover with a lid and bring to the boil, then boil for 5 minutes. Leave to cool for 15 minutes, lid on, to further infuse the flavour of the lemongrass into the liquid.

Strain the infused liquid into a bowl and stir in the honey. Leave until the liquid cools completely and the honey dissolves.

Pour the infusion into a jug or bottle, ideally one that has a lid or stopper. You can drink this as it is, just adding lime juice and serving over ice, but I like to add a touch of astringency by turning it into an iced tea. To do this, add the tea leaves to the liquid in the jug or bottle and put in the fridge. Leave to infuse for 4—5 hours or overnight, stirring or shaking occasionally. Strain off the tea leaves and add the lime juice. Taste the iced tea, adding more honey or lime juice if necessary. Serve chilled, over ice, with a slice of lime.

MOROCCAN MINT TEA COCKTAIL

Honey brings out the best in a classic Moroccan mint tea, here sweetened subtly with honey rather than the usual teeth-dissolving amounts of sugar.

The recipe comes from the chef Mark Hix, a champion of good British produce who also knows his way around Moorish flavours and moreish cocktails. Mark says you can top the cocktail with a sprinkle of ras el hanout, Morocco's classic spice blend.

SERVES 4–6

1 BUNCH OF MINT

350ML/12FL OZ JUST-BOILED WATER

2 TSP GOOD-QUALITY LOOSE-LEAF GREEN TEA

200ML/7FL OZ COLD WATER

3 TBSP CLEAR HONEY

ICE CUBES

100ML/3½FL OZ LEMON JUICE

150–200ML/5–7FL OZ VODKA, TO TASTE

Remove 4–6 sprigs of mint to decorate the cocktails. Put half of the rest of the mint in a teapot and pour in the just-boiled water. Leave to infuse for 5 minutes. Add the green tea and cold water. Leave to infuse for another 5 minutes, then strain off the liquid into a measuring jug. Stir in the honey, leave to cool, then chill in the fridge.

When you are ready to serve the cocktails, put the rest of the mint in a cocktail shaker or jug and smash it down, a process known as muddling. Bars and cocktail-swigging households will have a special piece of kit called a muddler; alternatively use the end of a thin rolling pin or another blunt implement.

Half-fill the cocktail shaker or jug with ice cubes, add the cold mint tea, the lemon juice and vodka and shake or stir vigorously for 20 seconds or so, then strain into serving glasses.

Top with the reserved mint sprigs and, if you like, sprinkle a little ras el hanout on top of the cocktails before serving.

AROUND THE WORLD IN 90 POTS

This is a guide to the most widely available honeys – and some of the rarer kinds – from around the world. As well as your local wildflower mixes, it is interesting to seek out honeys from other countries as they can provide a wide range of flavours; for example, the great honeys of New Zealand and the eucalyptus honeys of Australia come from a different flora from that of Europe or North America.

This extensive selection shows the range of colours, textures and flavours, but while some honeys are widely exported, others are only found close to where they are produced. When travelling look out for local honeys and try to have a taste. Whether monofloral or multifloral, bear in mind that honeys differ from place to place and year to year, depending on, among other things, the weather, details of climate, soil, production method and, most of all, the exact mixture of nectars gathered by the bees in any particular year.

ACACIA OR BLACK LOCUST

The false acacia (*Robinia pseudoacacia*), or black locust tree as it is called in the United States, provides good nectar for bees and makes honey that is high in fructose; it is therefore especially sweet-tasting and also remains runny rather than naturally crystallizing over time. This light, delicate honey is low in acidity and gently flavoured. Hungarian acacia honey is famous; it is also produced in quantity by Bulgaria and Romania, elsewhere in Europe and in the US.

AFRICAN FOREST, TROPICAL RAINFOREST OR JUNGLE

These multifloral honeys come from the vibrant mix of flowers in tropical rainforests or jungle. They are often dark, sometimes with powerful tastes and generally strongly aromatic, and can be harvested from indigenous or modern hives. The sales of such honeys help beekeepers in developing countries to found and run small businesses.

ALFALFA

Alfalfa's purple flowers cover large swathes of the United States, from the Midwest to the Pacific. The crop is used for feeding livestock. When it is left to bloom, the bees take advantage to produce a light honey that is good for everyday use. It is widely used in baked goods rather than being sold as a monofloral honey.

APPLE BLOSSOM OR ORCHARD

Springtime apple blossom makes light honey that can have green or reddish shades and is softly sweet. Apple trees are not great producers of nectar and the honey may well include nectar from other fruit tree blossom to produce an orchard honey.

ARBUTUS

Arbutus, or the strawberry tree, with its strawberry-like fruits, produces a valued medium-dark honey that is known for its medicinal properties and is relatively high in antioxidants. Beekeepers take their hives to be near this evergreen flowering shrub in Sardinia and Portugal, the main sources of this rare and special honey. A beguiling mixture of aromatic sweetness and bitter notes, it is known in Italy as corbezzolo honey, or as *miele amaro* (bitter honey).

ASPHODEL

A light honey with a slightly metallic, fennel flavour and a subtly fragrant finish that has been described as 'almond milk', this highly distinctive varietal is much praised by some honey-lovers, though not to everyone's taste. It comes from a herbaceous plant with large white flowers that grows in meadows in Sardinia.

ASTER

This light amber honey from the United States comes from late summer and early autumn flowers – some of the last flowers to feed the bees before the colony closes down for the winter – and has hints of thyme. It forms hard crystals over time. Sometimes aster nectar combines with that of another late-blooming plant, goldenrod.

AVOCADO

Rich and velvet-smooth, avocado honey is dark amber with a butter-caramel and malty aroma and taste that is much admired by honey-lovers. California, Mexico and Chile are the main producers. Good for using in chocolate or nutty puddings, for eating with fruit or cheese, for glazing and for using as a final flavour note in rich savoury dishes.

BALSAM

Himalayan balsam is an invasive plant with pink flowers that quickly spreads along riverbanks. Regarded as a rampant weed by some, it is loved by the bees, who use its nectar to produce a light honey that is subtly sweet and full of flavour. The honey is often part of wildflower mixes and beekeepers say it wins them prizes at honey shows when the balsam predominates in the mix.

BASSWOOD *see* LIME

BEAN OR FIELD BEAN

This amber honey is not often advertised as 'bean' as this sounds too savoury for a sweet substance. Nevertheless, honeybees and bumblebees pollinate the white flowers and gather nectar from field broad beans, such as those grown widely for animal feed in the UK. The nectar can be useful to feed bees during what is called the 'hungry gap' of June, when there are surprisingly fewer flowers between spring and high summer. The honey is popular with some tasters, though unusual to find commercially.

BEECHWOOD

A dark, honeydew honey with a malty flavour that is produced from beech forests in New Zealand and elsewhere; it is relatively antibacterial and high in antioxidants.

BELL HEATHER

Honey made from bell heather (*Erica cinerea*) is different to honey made from ling heather (*Calluna vulgaris*). More of a purply amber or 'port wine' colour, this medium honey is also fragrant with hints of geranium in the flavour. Other species of moorland heathers also make good honey – for example, in Spain and France. *See also* Heather.

BLACKBERRY

A light to medium amber with a good berry scent, this honey crystallizes quickly and quite coarsely. Major sources are the berry-growing areas of the north-western United States. In Europe it can be a nectar source that is part of summer honeys. Blackberries have a long flowering season, from June to August – you often seen bushes with both fruit and flowers on them – and this means plenty of food for the bees.

BLACKBUTT

A eucalyptus honey from Australia that has a reputation for being healthy. Dark, with flavours of molasses and stout. Infrequent flowering makes it rare.

BLACK BUTTON SAGE

Black button sage honey, the most renowned of the various sage honeys, comes from a perennial shrub that grows on the Californian coast and in the Sierra Nevada. Light in colour, it has a rich, layered flavour with hints of fig, menthol and butterscotch.

BLUEBERRY

An amber honey that is well rounded but also more tangy than other berry honeys, blueberry honey comes mainly from farms in New England and Michigan and also from wild blueberry bushes. Its butterscotch flavour, with green leaf and citrus notes, make it a good choice for drizzling on sweet dishes as well as for baking.

BLUE GUM

The blue gum is the floral emblem of Tasmania and is found here and in southern Australia. The cream-coloured flowers produce copious nectar for a medium honey with a slightly minty flavour.

BORAGE

A pale, delicate and sweetly fragrant honey. Borage is grown as a crop for the pharmaceutical industry for its gamma-linolenic acid, which is used as a health supplement, and as a result this varietal honey is now more widely available. If you grow borage at home you can both feed bees and enjoy eating the beautiful blue flowers. With their slightly cucumbery flavour these edible flowers are good not just in Pimm's and other drinks but also in honey-dressed salads.

BUCKWHEAT

Dark and strong, buckwheat honey is traditionally used in gingerbreads and central European baking and is a good choice for a strong, malty glaze or savoury dishes such as barbecue sauces. The fragrant white flowers produce nectar for a honey that is slightly spicy with a touch of molasses. This honey is comparatively rich in antioxidants and minerals and is used for medicinal purposes to soothe a tickly throat or for stomach-ache.

CAROB

The seed pods from this plant are sometimes used as a chocolate substitute and as a thickener by the food industry in foods such as ice cream. The flowers' nectar makes a prized golden or dark amber aromatic honey with a liquorice or coffee-like tang. Some tasters also detect a slight chocolate note. A gourmet honey, it is produced in small quantities on carob farms particularly in Spain and Portugal, with some production on Italian islands such as Sicily and Sardinia and also other countries, including Argentina.

CARROT BLOSSOM

A medium to dark amber honey, this is collected from hives that are placed in the fields of commercial carrot seed growers, for example in Oregon in the United States. It initially has a caramel flavour and then a slightly vegetal, musky, earthy tang in which you may find notes of pepper.

CATCLAW

A browny-yellow honey, or darker in colour if from a drier desert source, this comes from a plant with long clusters of yellow flowers and curved spiky thorns, hence the name. It is most commonly found in Arizona, Texas and southern California and in other parts of the United States, depending on rainfall. The honey is spicy and buttery with a strong follow-through that some find has notes of cough mixture and cinnamon.

CHESTNUT

One of the classic gourmet dark honeys, this honey from the blossom of the sweet chestnut tree (*Castanea sativa*) isn't to everyone's taste but others love its resinous, slightly bitter flavour and regard it as one of the best honeys in the world. The colour can vary from brown to near-black. High in minerals, and also tannin, it is used for health purposes. Italy and Spain are the main source of commercial honeys. A classic Tuscan dish is to drizzle chestnut honey over aged pecorino cheese, served alongside pears. This honey is different to that of the horse chestnut (*Aesculus hippocastanum*), the common ornamental tree. This is a good plant for bees, producing light honey that granulates smoothly and is part of multifloral honeys, and a red pollen that makes the bees look as if they've been sprinkled with brick dust.

CHIA

An Australian honey from a plant with a reputation as a superfood, this is a golden honey with nutty flavours.

CHRISTMAS BERRY

The Christmas berry shrub (*Schinus terebinthifolia*) is native to Brazil and a source of honey in some tropical countries such as Hawaii. A light to medium amber, it has touches of brown sugar in its flavour.

CLOVER

This sweet, light, mild yet aromatic honey is so widely popular that many think of it as the typical honey flavour. It crystallizes easily and is often found creamed or ready-granulated as a white, fine-grained honey; the liquid honey is a light gold. The wild clover that was once such a part of old-fashioned meadows is one of the most prolific nectar producers. Modern farming methods mean clover is no longer quite so widely grown, or cut for silage as it flowers. Nevertheless, clover is a widespread fodder crop in temperate countries, and is also useful for fixing nitrogen in the soil. Of the many agricultural clovers, white and yellow sweet clovers are still important for honey production along with alsike clover, red clover and crimson clover. Sweet clover honey is also known as melitot honey and is used in folk healing, particularly in Russia.

COFFEE

Eva Crane, who was the British queen bee of all matters honeybee, mentions in *A Book of Honey* that there is a characterful honey made when bees fly from orange groves to coffee plants; an intriguing combination. The plant's nectar is part of some of the honeys of Ethiopia that are championed by the Slow Food movement.

COTTON

The southern United States has great swathes of cotton fields that produce a light, mild, all-purpose honey with a slight tang that is useful for cooking as well as eating as it is.

CRANBERRY

This reddish medium honey with its berry-tartness comes from the northern United States. The plants may bloom for just two weeks and so it is a rarity sought out by US honey-lovers.

DANDELION

One of the most intensely yellow honeys, the egg-yolk glow of dandelion honey is a welcome sight in the kitchen. It granulates hard and unevenly, however, and is best to attack with a knife rather than a bendy spoon. Dandelions give honeybees plenty of nectar and pollen even in low temperatures during April and May when the colony is growing and needs feeding up.

DESERT WILDFLOWER

A multifloral honey of different colours from the southern United States and anywhere with a desert landscape, this honey is composed of a mixture of the nectar of plants that survive in dry conditions. A southern States mixture might include mesquite and cactus among the rich mix of blooms.

EUCALYPTUS

Australia's most important tree produces a wide variety of honeys from its many species. The trees flower sporadically — some only every two to eight years, or more — but can produce copious amounts of nectar. Red gum, marri, karri, jarrah, moort, swamp messmate, yellow box and wandoo are just some of the honeys that you find in farmers' markets and shops. There are more than 500 species of eucalyptus native to Australia and it also grows in California, Sardinia and elsewhere. The honeys can have a medicinal tang and some types, such as jarrah, are known for their actively healthy qualities.

FIREWEED OR ROSEBAY WILLOW HERB
A light honey with a fragrant sweetness that is more complex than many other light honeys. Wildflower honeys with plenty of fireweed can win prizes at shows. The bright purple-pink flowers of rosebay willow herb are seen throughout temperate zones of the northern hemisphere – often beside railways and on riverbanks – and the plant gets its American name from being the first to grow in an area that has been razed by flames.

FOREST OR FIRTREE
Honey made from honeydew, the secretions of aphids feeding on tree sap in pine, oak and beech forests. Honeydew honey is dark amber or near-black and has a strong, resinous flavour that works well in savoury dishes, including glazes. This love-it-or-hate-it honey can become addictive.

FYNBOS OR KAROO
Fynbos means 'fine bush' in Afrikaans and this perfumed multifloral honey is from an area in the Cape of South Africa, including the semi-desert landscape of the Karoo. Medium or dark in colour, the honey comes from nectar from the special flora of the natural shrubland and heathland in an especially biodiverse area that includes both mountain and coast.

GALLBERRY
This prized honey from the south-eastern US, especially Georgia, is a medium honey that is slow to granulate, distinctively perfumed and has a flavour with hints of spicy cinnamon. It comes from a low shrub of the holly family, also known as inkberry, reputed to be one of the best honey plants in the US. The Slow Food movement has put it in the Ark of Taste of special foods that should be championed.

GOLDENROD
A thick honey that crystallizes quickly, this has a strong aroma in the hive that mellows once harvested; it is often used as a baking or everyday table honey. Goldenrod has a long season and provides plenty of nectar for the bees. Its name describes both the golden colour of its tasselled blooms and the hue of the honey, which can vary from light to medium gold.

GREEK MOUNTAINSIDE
One of the most famous honeys in the world, Greek mountainside honey comes from a mixture of fragrant herbs, particularly marjoram, thyme and savoury. Dark or amber and with a flowing viscosity, this is a honey to drizzle over thick yogurt or good cheese.

GUAJILLO OR HUAJILLO

This wild desert bush is native to south-west Texas and Mexico and its blossoms produce nectar for a light, perfumed honey with hints of lavender. This is one of the honeys in the Slow Food's movement's Ark of Taste, which champions special foods worth preserving.

HAWTHORN

Hawthorn blossom varies greatly in its production of nectar but in good years it makes a dark amber honey that is prized for its slightly nutty flavour. Hawthorn can bloom at the same time as apple and the honey from their combined nectars is highly rated by beekeepers. Most commonly, hawthorn is part of wildflower multifloral honeys in the UK.

HEATHER

A superb honey, with commercial production particularly in Scotland and the north of England, Ireland, Spain and France. Beekeepers move their hives to catch the ling heather bloom that purples moorland and hillsides in the late summer. Some years there is plenty of honey and other years almost nothing. Heather honey is a foxy red that can vary in shade, with the best tending to be on the light to mid-amber side, indicating that the honey hasn't been heated too much. The slightly resinous, aromatic quality has all the strength of dark honeys but without being too assertive. Its gel-like consistency means that small bubbles are trapped in the honey, giving it a characteristic appearance, sometimes described as sparkling. Heather honey can have a higher water content than other honeys and may be heated by processors concerned about fermentation. But if heated too much, the honey loses its fabulous bouquet and quality. Because it is slightly more difficult to extract than other honeys, it can often be found in honeycomb form. Mead made from heather honey has been prized for many centuries.

HONEYDEW *see* FOREST OR FIRTREE IVY

Greenish-yellow ivy flowers are the last important source of nectar and pollen for honeybees late in the year and often part of autumn honeys, giving a greenish tinge and a strongly aromatic flavour that reminds some of menthol.

JARRAH

A medicinal eucalyptus honey from Western Australia, this can be measured for its active antimicrobial properties and given a 'Total Activity' (TA) rating. The jarrah is a magnificent tall tree that can live for 1,000 years; it flowers biennially if the temperature and rainfall are right. The reddish-amber honey is high in fructose and does not crystallize. Champions claim it to be as powerful as manuka honey and with a better flavour.

KARRI

Western Australia's karri eucalyptus flowers once every ten years; it produces a medium honey with a smooth texture and slightly smoky flavour.

KIAWE (HAWAIIAN WHITE)

This honey from Hawaii is light and rich, with menthol notes, and crystallizes quickly. Kiawe is a rare and prized gourmet honey, produced in small quantities from the greenish-yellow flower spikes of the kiawe tree; it has to be harvested by the beekeeper at just the right time.

KUDZU

Harvested in the south-eastern United States, kudzu is an unusual and rare honey with purplish hints and a slightly grape-like taste. The purple, maroon and pink flowers of the plant tend to be foraged by bees mostly in dry years, making this an occasional rather than regular honey.

LAVENDER

A light honey with production centred in France and Spain, this special fragrant honey often crystallizes from a runny gold to a near-white honey with a spreadable smooth texture. Subtly floral, it captures the essence of a sunny summer. This is the sort of pot you bring back from holiday and eat throughout the year to dream of sunshine.

LEATHERWOOD

A prized, piquant honey from Tasmania, this exotic, light amber honey is gathered from the white blossom of the leatherwood tree. It can granulate to be so hard it is cut up into cubes and wrapped in paper like a sweetie. Honey-lovers praise its spicy, distinctive and complex flavour, which Slow Food describes as having balsamic scents and fresh notes of citrus fruits.

LEHUA

A gourmet honey from Hawaii, this light honey crystallizes smoothly. Its delicate flavour has butterscotch notes. The bees gather nectar from beautiful flowers that look like scarlet firework bursts all over the native ohi'a tree.

LEMON BLOSSOM

A golden honey with a smooth, crystallized texture, this has a fresh, tangy yet mellow flavour with hints of citrus. You can sometimes find artisanal lemon blossom honey sold on tables next to lemon groves in Mediterranean countries.

LIME, LINDEN OR BASSWOOD

Light gold with a touch of green, this honey granulates smoothly and has a distinctive minty tang. The drooping flowers of lime trees (from the *Tilia* genus rather than the citrus tree) produce copious amounts of nectar — most of all when the temperature is around 20°C/68°F — a major source of honey, not least for urban beekeepers, since lime trees are common in roads, gardens and parks. The nectar is mostly secreted before midday and the bees work lime trees in the morning. Most strongly flavoured honeys are dark but lime honey is an exception, being light but with the kick of its slightly astringent minty flavour.

MACADAMIA

Macadamia trees are the winter-time nectar source for a prized medium honey with a nutty, caramel flavour and hints of tropical fruit. This rich, soft and thick honey comes from the nut orchards in Hawaii.

MANGO

This dense, dark amber honey has a subtle fruitiness and can crystallize easily. Part of Jamaican and other Caribbean multifloral honeys, it is sometimes produced as a monofloral honey near mango farms in tropical countries where honeybees – along with ants and other bees – pollinate the cascades of flowers.

MANUKA

Nectar from the usually white flowers of the manuka bush, native to New Zealand and parts of south-east Australia, produce this medium honey with a distinctive medicinal smell that is almost antiseptic. Before it came into favour as a health food (see *pages 33–34*) some New Zealand beekeepers used to leave manuka honey for the bees, as an inferior honey that they couldn't sell because of its overly strong taste. This special healthy honey now commands a premium price and is graded with a 'Unique Manuka Factor' (UMF) rating and other grading systems to show the potency of a particular pot. Manuka's high protein level makes it viscous in texture.

MARJORAM

Marjoram produces especially sweet nectar and is popular with honeybees. It is one of the herbs that form part of the medium to dark Greek mountainside honey.

MEADOWFOAM

A delicate, medium amber honey, this comes from plants that are native to Oregon and other West Coast states, where they are also cultivated for their seed oil. The flavour has notes of vanilla and marshmallows.

MESQUITE

This honey, from the flowers of a tree mainly found in northern Mexico and the south-western US, varies from light to amber and has an earthy or smoky flavour, appropriately enough, as the sweet-smelling wood is used to flavour barbecue and smoked foods. The honey is sought out by honey-lovers, who sometimes use it to replace brown sugar in dishes.

MINT

Mint honey comes from crops that are allowed to flower in order to get the essential oil for products such as chewing gum. Commercial mint production is mainly based in the US, in Oregon, Indiana, Idaho, Ohio and Michigan; and you can also find mint honey (*miele de menta*) in northern Italy. The honey's sharp aroma is echoed in its piquant menthol aftertaste.

NEEM

Highly esteemed in Indian ayurvedic medicine, this dark honey has a liquorice flavour and may be collected from wild beehives in forests where the neem tree produces its fragrant white flowers.

OAK

This honeydew honey comes from the bees feeding on aphid secretions in oak woods. A dark, smooth honey with a strongly aromatic flavour, this is a good choice to drizzle over cheese or to use as a glaze. The commercial product comes from places with beautiful oak woodlands such as Catalonia in Spain, Bulgaria and France.

OILSEED RAPE

Oilseed rape (OSR), grown for vegetable oil, is an important early nectar source for bees. The light honey crystallizes quickly and beekeepers have to extract it before it hardens in the comb. Some varieties of OSR produce honey with slightly cabbagey notes but most are super-sweet and granular.

ORANGE BLOSSOM

One of the most common and distinctive light honeys in the world, orange blossom honey has floral, citrus notes that link the honey directly to its source. Its sweet, attractive nature and reasonable price make orange blossom honey one of the most useful and delightful all-purpose honeys to have in the kitchen.

PALMETTO OR SAW PALMETTO

Bees gather nectar from the flowers of this tall tree that grows in the south-eastern United States. The honey varies from light to amber, is light in body and full of flavour, with a slightly woody, herbal quality.

PHACELIA

This genus of 200 plants, native to North and South America and introduced to Europe in the early nineteenth century, produces flowers that are such bee-magnets that it has been suggested that phacelia should be grown as a crop specifically to produce honey. Grow some in your garden and watch them quiver with bees. The amber honey granulates quickly.

POHUTUKAWA

A very pale New Zealand honey that crystallizes quickly and has a slightly salty flavour and silky texture. It comes from the coastal 'Christmas tree', which produces flaming red flowers that peak in December and gush with nectar.

PUMPKIN

A medium amber honey with a spicy and rich flavour, this is a good honey to use as a glaze or in spiced honey cakes. It is a by-product of bees pollinating commercial pumpkin and squash and is often part of multifloral honeys.

RASPBERRY

Unlike most fruits, which do not necessarily give their particular flavour to a honey, raspberry nectar imparts a slightly fruity, raspberry tang to its honey. This light, sweet honey also has floral notes and is a popular gourmet honey, good for both sweet and savoury dishes.

RATA

A New Zealand honey mostly from the South Island, this has a silky granulated texture and a floral, slightly menthol and salty flavour. It comes from a crimson-flowered plant related to the one that produces the prized lehua honey of Hawaii.

REWAREWA

From a red-flowered tree known as the New Zealand honeysuckle, rewarewa honey is a medium to dark amber with a slightly reddish tint and a rich and distinctive slightly burnt or malty flavour. It is a healthy honey, with anti-bacterial properties.

ROSEMARY

This medium honey, lighter when creamed, has floral, herbal notes and is part of the famous honey of Narbonne in the South of France (also containing thyme and other moorland flowers), which has been praised since the time of ancient Greece and Rome, and also comes from Spain, Italy and other European countries. In France it is called *Miel de romarin* and there is specialist production in Corbières in South East France.

SAFFLOWER

A honey that comes from fields of a plant grown in the United States for its cooking oil. The honey is dark and has a strong, even pungent, scent with a touch of spice and a flavour that some like and others find too powerful.

SAGE

From the western United States, sage honeys come plants of the *Salvia* family and tend to be light and softly fragrant. *See also* Black button sage.

SEA LAVENDER

A rare British honey that is well-flavoured, with a hint of bitterness; the nectar is gathered from purple marshland flowers on the East Anglian coast that bloom in August and early September. This is part of Slow Food UK's Forgotten Foods (UK Ark of Taste) programme.

SIDR OR JUJUBE

An elite honey from Yemen, this commands a very high price because of its status as a rare and healthy honey, especially in the Middle East. The nectar flow of the sidr tree is limited, making the honey especially scarce and pricey. It is a dark amber honey with earthy, butterscotch flavours.

SOURWOOD

This tree-flower honey from the southern and eastern US, most abundant in the Alleghenies and the Blue Ridge mountains, is one of the most prized in the country. The short and unpredictable summer bloom – about 25 days - makes it difficult to collect as a single varietal honey and all the more precious. A good crop may be gathered only once a decade. Varying from light to medium in colour, it has a delicately spiced flavour with notes of flowers and anise.

STAR THISTLE

A honey that could typify the title of Patience Gray's book, *Honey from a Weed*, this comes from a plant (*Centaurea solstitialis*) that is regarded as an annoyance by farmers in California and other states, not least because of its long taproot. However, it produces a light, subtle mild honey.

STRINGY BARK

An Australian eucalyptus honey from the nectar of several species of tree named after their fissured bark, this is a strong, dark honey with plenty of tangy flavour.

SULLA

Honey from the sulla or French honeysuckle (*Hedysarum coronarium*) is found especially in Tuscany, the south of Italy and Sicily, and is a light runny or white crystallized honey with a fine texture and mild, sweet flavour. Often part of wildflower honeys, it can sometimes be found as a varietal.

SUNFLOWER

A beautiful bright gold, this is a slightly minty, fresh-tasting honey that crystallizes quickly and is produced by beekeepers in southern Europe.

TAMARISK

A dark honey with malty, savoury, woody flavours; high in fructose, it is slow to granulate. Tamarisk, or salt cedar, is a plant that grows in dry places such as New Mexico, Texas and Mexico.

TAWARI

The beautiful white flowers of New Zealand's elegant tawari tree are largely bird-pollinated and their copious nectar is also used by honeybees to produce a light honey with a butterscotch taste. A high fructose level makes the honey taste especially sweet.

THISTLE OR CARDOON

If the bees are near a field of thistles, they can produce a honey that is said to be slightly spicy, woody or insistently floral, like geranium. One of Sardinia's famous honeys is from the plant *Galactites tomentosa*; the milk thistle, or Scotch thistle (*Silybum marianum*), is another source of thistle honey. Most thistle honey goes into wildflower honeys.

THYME

One of the most famous and loved medium or dark honeys in the world, this is a staple flavoursome honey in Italy and Greece. Greek thyme honey is delicate yet highly aromatic and rich in enzymes and flavour. *See also* Greek mountainside.

TRUFFLE

Strictly speaking, this is a flavoured honey. Slithers of truffle are put into light honey, often acacia, giving this Italian honey a strong and slightly feral tang that works brilliantly when drizzled over savoury foods such as cheese, adds an intriguing perfume to a salad dressing and turns a plate of mushrooms on sourdough toast into a sensational starter or lunch dish. While expensive, truffle honey is an easy way to make food special.

TUPELO

From a tree that grows in the wetlands of Florida and Georgia. The blossom flows with sweet nectar for just a short while and its high fructose content means it does not granulate. The flavour and light colour make it shine which is why Irish musician Van Morrison's love song and title track for his 1971 album *Tupelo Honey* has the lyric: 'She's as sweet as tupelo honey.'

ULMO

From an evergreen plant with white blossoms that grows in the Chilean rainforests, ulmo honey is a light yellow with fine crystals, with an aroma that is floral and then butterscotchy, with notes of jasmine and aniseed. An 'active' or medicinally significant honey.

URBAN HONEY

Beekeeping has returned to cities as people have become more aware of the need for bees. Tests show that urban honey is free from pollution. These honeys can be varied, derived from a mixture of flowers and trees from gardens and parks. Bees used to be part of urban life but became more rare and were even banned from New York City. They are now legal again after an 'overground' underground movement, led by beekeeper David Graves, gained publicity for rooftop hives. Other US cities have long championed urban bees and the number of beekeepers has soared in London and other cities. Paris even has a law, passed in 1895, stating that hives are allowed so long as they are five metres away from neighbours.

WILDFLOWER

Wildflower honeys vary from late spring to early summer and mid to late summer and then autumn. A mixture of whatever the bees can find, these can be magnificent in places where there is low-intensity farming and plenty of forage in hedges and fields, and in special wild areas such as the Alps and other highland pasture. In Italy it has the picturesque name of *millefiori* (a thousand flowers).

YELLOW BOX

A popular eucalyptus honey from Australia, this golden honey is high in fructose and therefore slow to granulate. The Latin name for the tree, *Eucalyptus melliodora*, means 'scent of honey', because of the smell of the feathery white flowers.

BIBLIOGRAPHY

Berenbaum, May (ed.) *Honey, I'm Homemade* University of Illinois Press, 2010

Bishop, Holley *Robbing the Bees* Free Press, New York, 2005

Bissell, Frances *The Scented Kitchen: Cooking with Flowers* Serif Books, London, 2007

Clifton, Claire *Edible Flowers* Bodley Head, London, 1983

Crane, Eva *A Book of Honey* Oxford University Press, 1980

Crane, Eva (ed.) *Honey: A Comprehensive Survey* Heinemann, London, 1975

Crane, Eva *The World History of Beekeeping and Honey Hunting* Duckworth, London, 1999

Dalby, Andrew and Sally Grainger *The Classical Cookbook* British Museum Press, London, 2000

Dixon, Luke *Bees and Honey* Northern Bee Books, Mytholmroyd, 2013

Dixon, Luke *Keeping Bees in Towns and Cities*, Timber Press, 2012

Ellis, *Hattie Sweetness & Light* Sceptre, London, 2004

Furness, Clara *Honey Wines and Beers* Northern Bee Books, Mytholmroyd, 1987

Goldman, Marcy *A Treasury of Jewish Holiday Baking* Whitecap Books, Canada, 2009

Goulson, Dave *A Sting in the Tail* Jonathan Cape, London, 2013

Gowing, Elizabeth *The Little Book of Honey* Elbow Publishing, 2012

Hill, Shaun *Cooking at the Merchant House* Conran Octopus, London, 2000

Kelly, Sarah *Festive Baking in Austria, Germany and Switzerland* Penguin, London, 1985

Kirk, W.D.J. and F.N. Howes *Plants for Bees* International Bee Research Association, Cardiff, 2012

Little, Maureen *The Bee Garden* How To Books, Oxford, 2011

Nordhaus, Hannah *The Beekeeper's Lament* Harper Collins, New York, 2011

Marchese, C. Marina and Kim Flottum *The Honey Connoisseur* Black Dog & Leventhal Publishers, New York, 2013

Munn, P. and R. Jones *Honey and Healing* International Bee Research Association, Cardiff, 2000

Passard, Alain, trs Alex Carlier *The Art of Cooking with Vegetables* Frances Lincoln, London, 2012

Poole, Bruce *Bruce's Cookbook* Collins, London, 2011

Preston, Claire *Bee* Reaktion Books, London, 2006

Ramírez, Juan Antonio *The Beehive Metaphor: From Gaudí to Le Corbusier* Reaktion Books, London, 2000

Salloum, Habeeb, Muna Salloum and Leila Salloum Elias *Sweet Delights from a Thousand and One Nights: the story of traditional Arab sweets* I B Tauris, London, 2000

Sinha, Anil Kishore *Anthropology of Sweetmeats* Gyan Publishing, India, 2000

Style, Sue *Honey, from hive to honeypot* Pavilion, London, 1992

Tulloh, Jojo *The Modern Peasant* Chatto & Windus, London, 2013

Wyndham Lewis, Sarah *Planting for Honeybees* Quadrille, 2018

HONEY TO BUY

Bee Raw (USA) www.beeraw.com
Varietal honeys, including interesting tasting
flights.

Berkshire Berries (Massachussetts, USA)
www.berkshireberries.com
Pioneers of New York City honey.

Bermondsey Street Bees (London, UK)
www.bermondseystreetbees.co.uk
London and country artisanal honeys and
information about plants and sustainability.

Brindisa (London, UK) www.brindisa.com
Great artisanal honeys from Spain.

Chain Bridge Honey Farm (near Berwick-on-
Tweed, UK)
www.chainbridgehoneyfarm.co.uk
Heather and wildflower honey from
Northumberland and
Scottish borders, plus charming visitor centre.

Coedcanlas (Wales, UK)
www.coedcanlas.bigcartel.com
Raw, artisanal honeys from Pembrokeshire.

Follow the Honey (Boston, USA)
www.followthehoney.com
Specialist honey shop and supporter of all
matters bees and honey.

Fortnum & Mason (London, UK)
www.fortnumandmason.com
Wide selection of UK and world honeys.

From Field and Flower (London, UK)
www.fromfieldandflower.co.uk
High quality honeys from around the world in
London's Borough Market.

Heather Bell Honeybees (Cornwall, UK)
www.cornwallhoney.com
Cornish wildflower honey and beekeeping
supplies.

La Maison du Miel (Paris, France)
www.maisondumiel.com
Beautiful honey specialist, founded in 1898.

Les Abeilles (Paris, France) www.lesabeilles.biz
Great honey shop on the Left Bank.

The Hive Honey shop (London, UK)
www.thehivehoneyshop.co.uk
Beekeeping courses and a wide range of honeys.

The House of Honey (Australia)
www.thehouseofhoney.com.au
Raw honey from Western Australia

Ogilvy's Honey www.ogilvys.com
Ex-honey trader and chef with worldwide
selection, including honeys from New Zealand.

La Pecheronza (Puglia, Italy)
www.pugliamiele.com
Monofloral honeys including coriander
and cherry.

Seggiano www.seggiano.com
Top artisanal Italian honeys.

Wild About Honey www.wildabouthoney.co.uk
Special raw Portuguese artisanal honeys.

HONEY AND BEES
INFORMATION ONLINE

www.100daysofhoney.wordpress.com
Beekeeper and author Elizabeth Gowing's blog
of honey recipes.

www.albomiele.it
Honey tasting courses at the Bologna-
based register of Experts in the Sensory
Analysis of Honey.

www.americanhoneytastingsociety.com
US honey tasting and evaluation

www.beedata.com/localhoney
Database of UK honeys and beekeepers.

www.beesabroad.org.uk and
www.beesfordevelopment.org
Promote beekeeping in developing countries.

www.bbka.org.uk
British Beekeeping Association, with a directory
of beekeeping groups.

www.bwars.com
The UK Bees, Ants and Wasps Recording Society
for monitoring all species of bees.

www.honeyshow.co.uk
Annual UK honey show held in the autumn.

www.groovycart.co.uk/beebooks
Northern Bee Books, specialists in bee and honey
books and publications.

www.honeytraveler.com
Outstanding website about honeys around
the world.

www.honey.com
US National Honey Board website with
information, recipes and producers.

www.ibra.org.uk
International Bee Research Association, an
important source of information about bees
and honey.

www.jarrahhoneyinfo.com
Information about the health benefits of
Australia's jarrah honey.

www.llgc.org.uk
National Library of Wales, home of the Eva
Crane/IBRA collection
of bee and honey books and publications.

www.manukahoney.co.uk
Information about the health benefits of New
Zealand's manuka honey.

www.mazercup.com
US annual mead competition.

www.naturalbeekeepingtrust.org
Promotes sustainable beekeeping.

www.rosybee.com
Specialist nursery for bee plants.

www.slowfood.co.uk and www.slowfoodusa.org
Promotes good, clean and fair food,
including honey.

www.sussex.ac.uk/lasi
The Laboratory of Apiculture and Social Insects
(LASI) at Sussex University, largest UK research
group looking at bees.

www.urbanbeekeeping.co.uk and
www.urbanbees.co.uk
Promote urban beekeeping.

SPECIALIST PRODUCERS

argan oil www.belazu.com

edible flowers www.maddocksfarmorganics.
co.uk

fennel pollen www.globalharvestdirect.com

honey beers www.hiverbeers.com

lavender sugar www.steenbergs.co.uk

mead Ninemaidens Mead
www.cornwallsolar.co.uk

AUTHOR'S ACKNOWLEDGEMENTS

My thanks first of all go to the Robson family at the wonderful Chain Bridge Honey Farm near Berwick-on-Tweed, Northumberland. They inspired me to write about honey in the first place and have continued to offer inspiration and help over the last ten years, from gorgeous lip balm and heather honeycomb to answers to questions.

Thanks to the International Bee Research Association (IBRA), a great organization with an astonishing bee library now based at the National Library of Wales in Aberystwyth.

Thanks to James Hamill at the Hive Honey Shop, London SW11; David Perkins and all those working at Roots and Shoots, an environmental education programme in Lambeth, south London; the London Beekeepers Association; The Bee Team at the Lancaster London Hotel; Jojo Tulloh; Susan Wilmot; everyone I spoke to at the excellent National Honey Show; Shamus Ogilvy of Ogilvy's Honey; Richard and Charlotte Dunne at Ashley Church of England Primary School in Surrey; Tim Baker at Charlton Manor Primary School in south-east London; Dr Luke Dixon of Urban Beekeeping in London, Stuart Roberts at Reading University, Jeroimo Villas-Bôas, an expert in stingless bees, and all the other beekeepers who have kindly helped me with my queries over the years. Thanks to Belazu, Luscombe Organic Drinks and Marks & Spencer for fab honey travels.

For further honey travels and discussions after the first edition of this book, I'd like to thank Francesco Colafemmina of Le Pecheronza in Puglia, who taught me much about the world of monofloral Italian honeys. Gian Luigi Marcazzan of the Register of Experts in the Sensory Analysis of Honey in Bologna, Lucia Piana, Daniela Carretto and Beatrice Monacelli shared their experience of how to train as a honey taster for a feature in the *Financial Times*. Alison Knox brought me to the Nottinghamshire Beekeepers Association, Amy Shelton to her beautiful Princesshay Honeyflow Lightbox in Exeter and Kew Gardens to The Hive. A big buzz from Clare Densley at Buckfast Abbey, Sarah Wyndham Lewis and Dale Gibson of Bermondsey Street Bees, Hannah Rhodes of Hiver honey beers and bee author Helen Jukes. Everyone should seek out Eric Tourneret's astonishing photographs of bees and beekeepers (www.thehoneybeesphotographer.com).

At Pavilion Books, great thanks to Rebecca Spry for getting the project going and general buzz, and also Polly Powell, Fiona Holman, Georgina Hewitt and Claire Marshall. Thanks to Katie Cowan and Stephanie Milner for coming back for another spoonful. Big thanks to Maggie Ramsay for her excellent copy edit and to the incomparable Miranda Harvey for the beautiful page design. For the photographs, I couldn't have been luckier or happier with the team. Thank you Maja Smend, Sunil Vijayakar, Lucy Harvey, Natalie Thomson and Xenia von Oswald. Maja – you captured the beautiful and charming glow of honey with your lyrical mastery of light. Thanks to Laura Hynd for searching out photographs of bees and blooms. Thanks to Jan Billington at Maddocks Farm Organics who provided beautiful edible flowers for the photography sessions. Thanks are ever due to my agent, Georgina Capel at Capel and Land.

Over fifteen years of honey gathering, I want to thank my parents, Roger and Margaret Ellis, and to the many friends who have brought me pots of honey and honey-related photographs from their travels, especially Patti, Monica, Debs, Tricia and Andrew. My love and thanks to my husband Tim Neilson for his great love and support throughout the writing of this book, not least, as ever, for the washing up of (very) sticky pans.

PICTURE CREDITS

INDEX

HEATHER HONEY

WILD HERB & THYME

ROSEMARY

CHESTNUT

LIME BLOSS

ORANGE BLOSSOM

ENGLISH AUTUMN HONEY

SUNFLOWER

ACACIA

MIEL DU BUGEY (FRENCH WILDFLOWER HONEY)

LAVENDER

PYRENEES MOUNTAIN & OREGANO

Hattie Ellis was shortlisted for both the André Simon Awards Food Book of the Year and the Guild of Food Writers Awards Best Cookery Book for *Spoonfuls of Honey* in 2014. In 2013, Hattie won two Guild of Food Writers Awards for *What to Eat?* in the Food Book Award and the Miriam Polunin Award for Work on Healthy Eating categories. She is also the author of *Sweetness & Light: The Mysterious History of the Honeybee*, *The One Pot Cook*, *10 Chewy Questions About Food*, *Planet Chicken* and *Best of British Fish*. Hattie has written about bees and honey in numerous publications, including *The Times Magazine*, *Telegraph Magazine* and weekend sections, *FT Weekend*, *delicious.*, *Kew Magazine* and *The Field*.